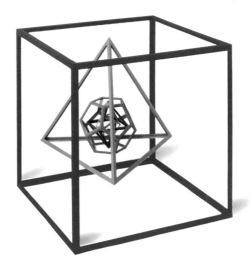

TIME &
SPACE

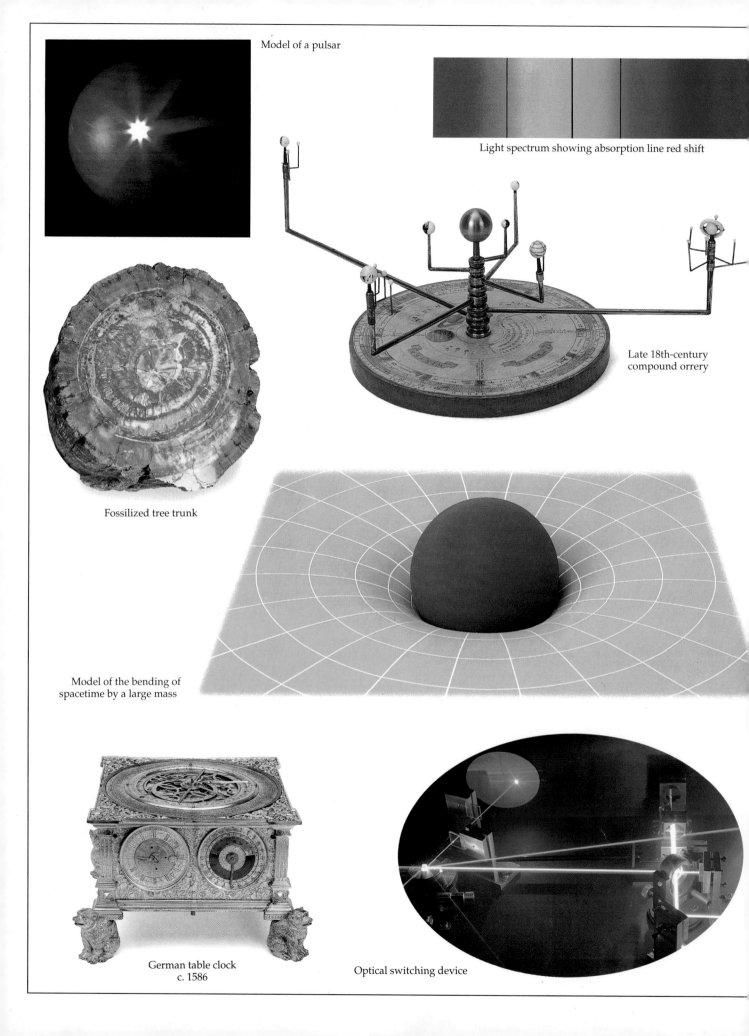

Model of a pulsar

Light spectrum showing absorption line red shift

Late 18th-century
compound orrery

Fossilized tree trunk

Model of the bending of
spacetime by a large mass

German table clock
c. 1586

Optical switching device

EYEWITNESS ● SCIENCE

Refracting telescope c. 1780

TIME & SPACE

Written by
JOHN AND MARY GRIBBIN

Celestial globe

Model of
the planet
Jupiter

Time machine
made of a twisted
black hole

Rocket using the
time machine

18th-century sextant

Harrison's
chronometer, H4

Curved geometry around the globe

DORLING KINDERSLEY
LONDON ● NEW YORK ● STUTTGART

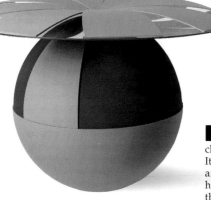

Curved, non-Euclidean geometry

DK

A DORLING KINDERSLEY BOOK

NOTE TO PARENTS AND TEACHERS
The **Eyewitness Science** series encourages children to observe and question the world around them. It will help families to answer their questions about why and how things work – from daily occurrences in the home to the mysteries of space. By regularly "looking things up" in these books, parents can promote reading for information every day.

At school, these books are a valuable resource. Teachers will find them especially useful for topic work in many subjects and can use the experiments and demonstrations in the books as inspiration for classroom activities and projects. **Eyewitness Science** titles are also ideal reference books, providing a wealth of information about all areas of science within the curriculum.

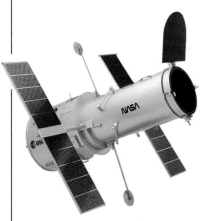

Hubble space telescope

Project Editor Liz Wheeler
Art Editor Ron Stobbart
Designer Helen Diplock
Production Adrian Gathercole
Picture Research Deborah Pownall
Managing Editor Josephine Buchanan
Managing Art Editor Lynne Brown
Editorial Consultant Professor Heather Couper

Nautical log

This Eyewitness ®/™ Science book
first published in Great Britain in 1994 by
Dorling Kindersley Limited, 9 Henrietta Street,
London WC2E 8PS

A CIP catalogue record for this book is available
from the British Library

ISBN 0 7513 1046 8

French pendulum clock, c. 1680

Reproduced by Colourscan, Singapore
Printed in Singapore by Toppan Printing Co. (S) Pte Ltd

Slave clock

Contents

Schrödinger's cat lives

Space and time

SPACE AND TIME SEEM TO OUR SENSES to be two very different things. "Space" is what we move around in, in three dimensions: left and right, forwards and backwards, up and down. We see things spread out around us in space, birds flying through the air, cows walking around in a field. We can move around in space from one place to another. But time is less controllable. We seem to "move" through – or with – time, whether we like it or not. This motion is always in the same direction, from the past to the future, and always at the same speed, 24 hours per day. We cannot go back for a second look at the past, nor can we jump forward for a preview of the future. Investigation in this century has led to scientists describing space and time in the same mathematical language, as one entity called spacetime. They describe time as the fourth dimension. Although we cannot "see" time spread out like space, equations like Pythagoras' theorem (p. 14) work in four mathematical dimensions to explain how things move.

RELATIVELY PROGRESSIVE THINKER
Isaac Newton (1642-1727) realized that motion is relative – a person walks at 7 km/h (4 mph) relative to the ground and the Earth moves through space relative to the Sun – but he believed there was an "absolute space" and an "absolute time" against which all positions and motion could ultimately be measured. In 1665 and 1666 Newton made several mathematical breakthroughs, including the binomial theorem, and calculus. He became Professor of Mathematics in Cambridge when he was only 26.

This place on the river represents 8 o'clock in time

RIVER OF TIME
One of the most common descriptions of time is like a river, an "ever-rolling stream" carrying us along for the ride. We move from the past to the future as unrelentingly as a river flows down from the mountains towards the sea. We think of a clock sitting in one place while ticking away the time. To take this a stage further, each time on the clock could correspond to a different "place" along the river – along the fourth dimension. At any point along the river, it is possible to remember events in the past, "upstream", but not what lies around the next bend, in the future.

MEASURING SPACE
Space and time are measured differently. Time is measured as if a tape measure is being unrolled, noting each marker as it passed – 1 metre (1 yard), 2 metres, 3 metres, and so on. Neither "end" of the measure can be seen, only the passing of the "markers". The only choice is how big an interval to take. Space is different because both ends of a space measurement can be seen at the same time, and a certain length of ruler carried around. A length of time cannot be carried around like a length of string.

Late 19th-century surveyor's tape measure

MEASURING TIME
We use clocks and watches to tell us when to get up and go to bed, when to eat our meals, when to go to work or school, and when to come home. We can measure time in fractions of a second, using a stopwatch, or count time in billions of years, back to the birth of the Universe. We cannot control time. If anything, time controls us. Humankind has always been controlled by time. The changing seasons dictate hunting and farming. The pattern of day and night determines when we are asleep and when we are awake. This has been true throughout human history. Nowadays we let time tell us what to do not just day by day, but hour by hour, or second by second.

SEKONDA
16 JEWELS

Late 20th-century surveyor's folding rule

MOVING FORWARD IN TIME

The most important difference between time and space is that we only move one way through time. There is a fixed "arrow of time" that determines everyday activities, such as baking. After the cook has finished, they can leave the kitchen, then change their mind and come back, travelling to and fro through space. But it would be no use having second thoughts about the baking. The ingredients could not be unscrambled – that would mean travelling back in time.

The ingredients of a chocolate soufflé

HUMANS INVESTIGATE OUTER SPACE

The word "space" conjures up an image of outer space. We live on one small planet, circling an ordinary star which is just one of 100 billion stars in the Milky Way galaxy. Our galaxy is one among many millions in the Universe. Some objects seen by astronomical telescopes are so remote that their light has spent 10 billion years on its journey to us. Motion is a marriage of time and space, so the time light spends on its journey is a measure of distance. One light year is the distance light can travel in a year, so the most distant objects are about 10 billion light years away.

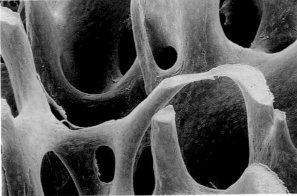

Combining the ingredients

HUMAN'S INNER SPACE

Space also extends inwards, to the world of the very small. In our bodies, the "solid" bones are partly made up of a fine lacework structure, (magnified in the picture above), adapted to support the body's weight. Some bone is more dense, but there are channels carrying blood and nerves through all bone, maintaining it as a living three-dimensional structure.

Pouring the mix into the baking dish

Rivers always flow in one direction, like time

This place on the river represents 8.20 in time

GOING UNDERGROUND

A lot of time is spent moving about in two dimensions, walking across the floor of a room, or travelling from one place to another by road. Within a building, people can also move about in three dimensions, sometimes travelling up and down stairs between floors, as well as moving around the floors. In more complicated structures, like an underground transport system, as much effort is devoted to movement in the third dimension as in the first two. Travelling up and down in space becomes as easy as moving from left to right.

The baked soufflé is a one-way trip

River has flowed through space, as time has ticked away on the clock

Ideas of the ancients

Arrow flies at 10 times the speed of the runner

Warrior sets off when he sees the archer about to take aim

Eand L GREEK THINKERS DESCRIBED THE WORLD as a flat disc, circled by a river. Others thought the Earth might be square – ancient Peruvians said it was shaped like a box, with a ridged "roof" where the gods lived. From earliest days, people wondered what holds the Earth up. Even after they realized that the Earth was spherical, the Greeks imagined that it was held up by the giant Atlas. In Mexico, Aztec people thought the heavens were held up by four gods. These ideas sound strange to us, because we are brought up on the idea that the Earth is a round ball floating in empty space, but they are all common sense ideas based on everyday experience. The ancients realized that human senses are inadequate to understand the Universe, but they did not have anything better to go on. Time also fascinated ancient thinkers. Some Greeks saw time as the ultimate judge, who would always discover and avenge any injustice. Some cultures, including the Chinese, Hindus, and some Greek philosophers, imagined time moving in alternating rhythms.

THE PARADOXES OF ZENO

The Greek philosopher Zeno of Elea lived in southern Italy from about 495 BC to about 430 BC. He made a series of statements, which are known as the "paradoxes of Zeno", designed to show that the human understanding of motion and time were inadequate. Plato (427-347 BC) realized that one way to solve the puzzles is if time behaves like space. Another way is if we accept that time cannot be chopped up into ever smaller pieces. Both these ideas are now an integral part of modern physics – the theory of relativity (pp. 34-35), and the quantum theory (pp. 52-53).

A GOD OF TIME

The ancient god of time, Chronos, appeared in the mythology of several civilizations around the Middle East. Both the Iranian god of time and the Greek Chronos were shown as winged serpents. In Roman times, Chronos kept his Greek name but was depicted in more human form, as in this marble statue from the 2nd century AD. Pythagoras (p. 14) in the 6th century BC, described the god Chronos as the "soul" of the Universe. The ancients had difficulty in understanding the flow of time, and created gods and myths to help explain it.

RACE AGAINST TIME

Zeno's paradoxes look at the relationship between time and space. His most famous puzzle "proves" that a tortoise given a head start can never be overtaken by any runner, even the fastest. This was designed to prove that the slower mover will never be passed by the swifter in a race. The same reasoning says that a warrior can always stay ahead of an arrow that is flying after him. Zeno's "arrow" paradox attempted to prove that a moving object is actually at rest. The puzzles depend on imagining that we can stop the action at any moment, like pressing the pause button on a video, to check up on how things are.

1 ON YOUR MARKS
When the Greek warrior sees an archer about to fire an arrow at him, according to Zeno all he has to do is turn around, and keep running at a steady speed. Imagine that the arrow is fired from 100 metres (110 yards) away just as he starts running, and that the arrow can fly ten times as fast as he can run. The idea relies on there being an infinite number of moments on "pause".

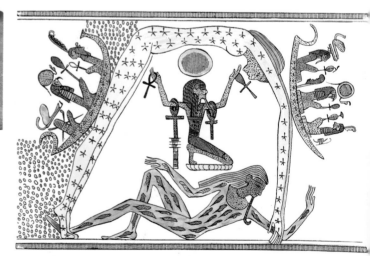

EGYPTIAN RULERS OF THE EARTH AND THE HEAVENS

This papyrus illustration was drawn in Egypt about 3,000 years ago. It shows Shu, the air god, separating his children Geb, the Earth god, and Nut, the sky goddess. The Sun god, Ra, and Moon god, Thoth, are carried across the sky in boats above Nut's starry body. The Egyptians believed that one day everything would dissolve into chaos, and a new creation would follow. This idea of a cycle of time was very common in many cultures from around the world. Some religions and philosophies still contain this idea – and it even turns up in the Big Bang theory (pp. 60-61).

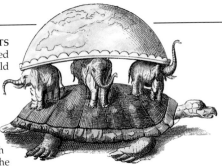

INDIAN OBSERVATORY
All over the world people began to use scientific observation to develop theories about space. This ancient open-air observatory was built at Jaipur, the capital of the Indian state of Rajasthan. It has stone structures used to measure the positions of the Sun, Moon, planets, and stars.

INDIAN ELEPHANTS
Different cultures formed their own views of the world around them. This Hindu engraving shows the Earth as a hemisphere supported on the backs of four elephants. They stand on the back of a giant turtle, which swims through the universal ocean. The elephants rotate on the turtle, reflecting the changing seasons. This world has a distinct "edge", over which the unwary sailor could fall.

Arrow flies 100 m, reaching the point the warrior left

Warrior has moved a further 10 m on

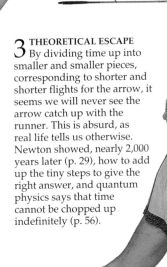

MAYAN OBSERVATORY
The Mayan civilization of Central America flourished between AD 600 and 900. The Mayans seem to have been more obsessed by time than any other ancient civilization. They·had a very accurate calendar, and studied the heavens from observatories like this one at Chichen Itza. Their calendar (p. 38) was actually more accurate than the one we use today, "losing" only one day in 5,000 years. Although modern Mayans are mostly Christians, the old cosmology has been absorbed into their religion. God is associated with the Sun, Mary with the Moon, and Christian saints with old Indian spirits.

2 KEEP ON RUNNING
By the time the arrow has flown the 100 m (110 yd) to the place where the runner started, the runner has had time to move on by 10 m (11 yd). He won't even be out of breath, and can keep running in the same direction at the same steady speed. By the time the arrow crosses the next 10 m (11 yd), the runner has had time to move on by another metre (1.1 yd). By the timed it crosses that metre, he has moved on by 10 cm (4 in), and so on. For every advance that the arrow makes, the runner advances too.

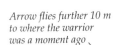
Arrow flies further 10 m to where the warrior was a moment ago

Warrior has run another metre

3 THEORETICAL ESCAPE
By dividing time up into smaller and smaller pieces, corresponding to shorter and shorter flights for the arrow, it seems we will never see the arrow catch up with the runner. This is absurd, as real life tells us otherwise. Newton showed, nearly 2,000 years later (p. 29), how to add up the tiny steps to give the right answer, and quantum physics says that time cannot be chopped up indefinitely (p. 56).

TIME STANDING STILL
Until this century, time and space have always been seen as two entirely different things. The Earth exists in space, as part of the physical Universe. Like everything else, it is carried along inexorably by time, as shown in this French engraving. We can stand still on the surface of the Earth, but we can never stand still in time. This is one reason why Zeno's paradoxes do not work in the real world. Time could be laid out like a map, with the future already determined, or we could have a choice about where we are going.

According to Zeno's paradox, the warrior will escape

Made to measure

SPACE IS SPREAD OUT ALL AROUND US. It is easy to see how much space there is between our steps when we take a pace, or how much space there is between the ends of the fingers on our two hands when we stretch out our arms. Early units of length were based on "measurements" like this – human paces, arm spans, feet, and thumbs. Measuring space became important when people started to divide up the land into fields. Farmers wanted to be sure they had as much land as they were entitled to. At first, people were happy to pace out the sides of a field. Then they gradually became more fussy, and the units of length had to be standardized, so that the farmer with short legs didn't end up with less land than the farmer with long legs. The standardized units were still based on the dimensions of a human body. However, one particular person's measurements, usually those of the King or Emperor, were made into rulers representing the standard length. Copies would be made for general use. Later, standards were based on the dimensions of the Earth. Even astronomers use units that are based on the sizes of relatively nearby things. They measure distances across the Solar System in terms of the astronomical unit, which is the average distance from the Earth to the Sun.

SAND CASTLES AND SCULPTURES
People have been good at measuring space since ancient times. This "sand sculpture", known as the Candelabra, was drawn in the desert near the Paracas coast of southern Peru, long before Christopher Columbus crossed the Atlantic Ocean. The pattern could only have been made by people who could make large, accurate measurements. They must have made a smaller version of the design first, like the Egyptian drawing opposite, and then copied it on to the desert. In other places, patterns made of long, straight lines were made by clearing stones from paths across the desert.

DIVIDING UP THE LAND
Accurate measurements became more important as farming became more sophisticated, particularly when nomadic tribes began to settle in one area for good. This French illustration from the 15th century shows peasants at work on a feudal estate. Landowners wanted to be able to measure their land accurately, and make sure their neighbours had not "stolen" any land. Farmers had to plan how much of each crop to plant in the fields, so they needed to know the size of the fields. Tax gatherers would base their assessment of a landowner's wealth on the amount of land the farmer owned. All of this required standard units of measurement.

MAN OF TRIANGLES
Pythagoras lived in the 6th century BC. Born on the island of Samos in Greece, he travelled to southern Italy and became the leader of a religious and philosophical community. Its members obeyed strange rules, including abstinence from eating beans and refusing to allow swallows to nest under the roof. They discovered and proved many important mathematical rules, including the famous theorem about right-angled triangles.

Hypotenuse, 5 units long

25 squares, made up of 9 plus 16 squares

FAMOUS THEOREM
Pythagoras' theorem says that the area of the square drawn on the hypotenuse (the side opposite the right angle) of a right-angled triangle is equal to the combined area of the squares on the other two sides. Although Pythagoras was interested in the numbers themselves, this kind of practical mathematics could be used to relate different areas of land. Almost 2,500 years after he died, mathematicians discovered that his rule is the key to making measurements in curved spacetime, using relativity theory (pp. 42-43).

Side 4 units long gives 16 squares

Right angle

Side 3 units long gives 9 squares

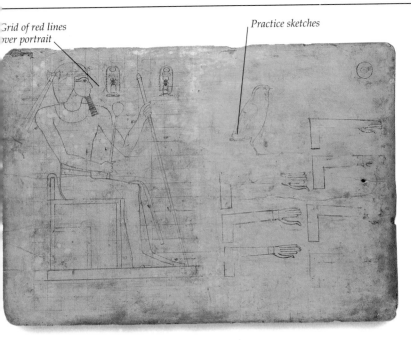

Grid of red lines over portrait

Practice sketches

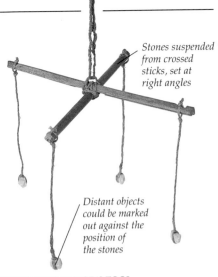

Stones suspended from crossed sticks, set at right angles

Distant objects could be marked out against the position of the stones

ARTIST'S DRAWING BOARD

Artists and architects also need to measure things accurately. This Egyptian drawing board from the 15th century BC is marked up with a grid. On the grid is a figure, possibly King Tuthmosis III, drawn with great technical skill according to the Egyptian convention of proportions. The artist could use this to "translate" his drawing into a large wall painting, or even a statue, while keeping the dimensions in the correct proportion to one another. No measurements of the actual lengths need be involved, but the relative sizes of different features, such as hands and arms, are used. The trick can also be used in reverse. If an artist holds up a frame containing a grid of strings he or she can easily copy the proportions of a scene on to paper.

ANCIENT SURVEYOR'S TOOL

Measuring angles is just as important as measuring lengths when surveying space. On a flat landscape, the ancient Egyptians could use this groma to line up directions to distant objects, and then scratch out the lines of sight on the ground. Alternatively, the angles could be marked out using the groma, and incorporated into buildings, such as pyramids. The technique worked well for the all-important right angles.

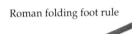

Roman folding foot rule

BUILDING ON A GRAND SCALE

Using equipment like the drawing board, the ancient Egyptians could make very accurate statues on a very large scale. These ones sit at the entrance to the great temple at Abu Simbel. They are perfectly in proportion, although they tower over the tourist standing in front of them. Without accurate means of measurement, the ancient Egyptians could never have built so well. The 20 m (60 ft) high statues were carved out of a sandstone cliff, and show Rameses II, who lived from about 1304 to 1237 BC.

Roman plumb line

ANCIENT ROMAN FEET

This bronze Roman foot rule folds up so that the carpenter can carry it around. One Roman foot was 296 mm (11.6 in) long, shorter than the modern imperial foot, and divided into 12 inches. A simple bronze weight on a string, or plumb line, was all that a Roman builder needed to give a perfect vertical line and make sure his walls were upright. With equipment like this, simple rules like Pythagoras' theorem, patience, and lots of work, structures like the Colosseum in Rome could be built. The Roman system was adapted from Greek standards of measurement, and the Greeks based their system on the Egyptian system. The Greeks used a unit called the finger, 19.3 mm (0.8 in), and 26 fingers made an Olympic cubit, copying the division of Egyptian cubits into digits. The modern inch is really the Egyptian digit in disguise!

AT A ROYAL ARM'S LENGTH

The Egyptian cubit was the standard unit of measurement for thousands of years. It was first devised around 3000 BC. This is part of an Egyptian cubit stick from about 600 BC, just before the time of Pythagoras. It is made of schist rock. One cubit is the length of an arm from elbow to extended finger tips. The standard royal cubit was made of black granite and kept in the royal palace. The royal cubit was 524 mm (20.5 in) long. It was subdivided into 28 digits. Sticks like this were made the same length as the master royal cubit and carried to other parts of the country.

REVOLUTIONARY MEASUREMENT

The standard metre was defined in 1799, by French revolutionaries, in terms of the size of the Earth. It was fixed as the distance from the North Pole to the equator, divided by 10 million. The standard metre above is made of platinum and iridium, and used to be kept under lock and key in Paris. As the metric system spread around the world, copies of the standard metre were made, so that they could be taken where they were needed, like the copies of the standard cubit made in Egyptian times. The metric system itself has now been refined, with a new, more precise, definition of the metre (p. 54).

Exploring the globe

MODERN CIVILIZATION HAS ITS ROOTS in the region around the Eastern Mediterranean Sea. Five hundred years before the birth of Christ, the world was thought to be a circular disc, with the Mediterranean more or less at its centre, surrounded by the land of Europe and what we now know to be the northern parts of Africa, India, and Asia, all in turn surrounded by an ultimate world ocean. About the year AD 1000, Norse voyagers had reached North America, but thought it was just another small island, not realizing its extent. Explorers gradually pushed back the frontiers of geographical knowledge, round the coastlines of Africa and India, and northward to Britain, and beyond. Greek philosophers were the first to suggest that the Earth is round, and by the time of Christopher Columbus this was well understood by educated people. The surprise discovery of Columbus's voyage in 1492 was not that the world is round, but that there is another continent in the sea between Europe and Asia. Columbus never knew that he had failed to reach Asia.

GREEK THEORIST
Anaxagoras was born in Greece around 500 BC, and was the first to explain that eclipses occur when the Moon passes in front of the Sun. He also suggested that all matter is made of infinitely divided substance (pp. 58-59). He calculated the Sun must be 56 km (35 miles) across and 6,400 km (4,000 miles) away from the Earth. His calculation was correct, but based on the mistaken assumption that the Earth is flat. Two centuries later, Eratosthenes used the same evidence to calculate the diameter of the spherical Earth.

ROSELLI CHART
This map, drawn by Francesco Rosselli around 1508, is one of the first to show the Americas. It accurately represents Europe and Africa from European explorations, but some of the vague eastern geography dates back to the ancient Greeks such as Ptolemy. South America is shown as an island, and Newfoundland and Labrador are joined to Asia. Names Columbus gave to places in Central America are shown on south-east Asia, following his belief that he had sailed to Indo-China on his fourth voyage.

MISTAKEN CALCULATION
At Syene near Aswan, at noon on the year's longest day, the Sun is directly overhead, so that the Sun's rays make a right angle as they fall on the Earth. At exactly the same time at the Nile Delta, the angle of the Sun's elevation is a bit smaller, so that the rays make an acute angle. If the Earth is flat, the geometry of right-angled triangles suggests that the Sun is 6,400 km (4,000 miles) overhead. However, as the Earth is round, the same measurements say that it is actually the radius of the Earth that is 6,400 km (4,000 miles).

Distance to Sun may apparently be calculated

Horizontal distance is known

Angle can be measured

Parallel rays from the Sun

Angle can be measured

Radius of the Earth calculated

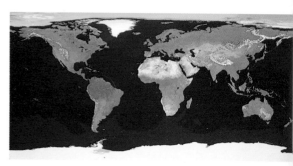

WORLD PANORAMA
Thousands of separate photographs taken from a satellite have been combined to make this cloud-free view of the Earth from space, which shows how accurate the Roselli map is. The image is centred over Africa, squashing the polar regions.

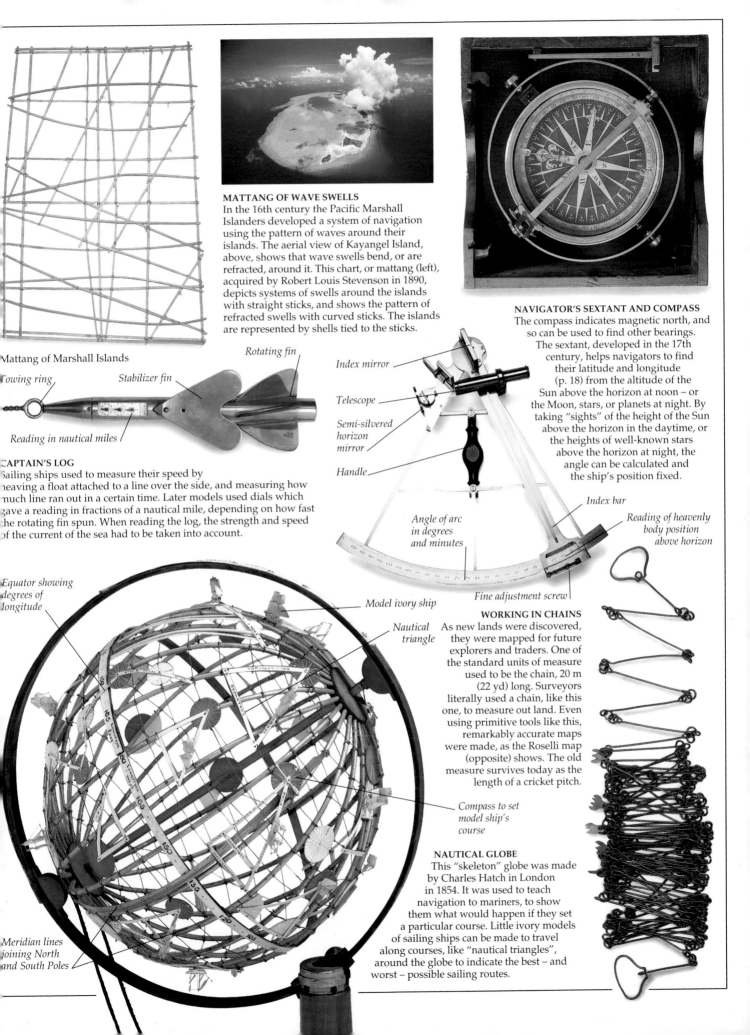

Mattang of Marshall Islands

MATTANG OF WAVE SWELLS
In the 16th century the Pacific Marshall Islanders developed a system of navigation using the pattern of waves around their islands. The aerial view of Kayangel Island, above, shows that wave swells bend, or are refracted, around it. This chart, or mattang (left), acquired by Robert Louis Stevenson in 1890, depicts systems of swells around the islands with straight sticks, and shows the pattern of refracted swells with curved sticks. The islands are represented by shells tied to the sticks.

NAVIGATOR'S SEXTANT AND COMPASS
The compass indicates magnetic north, and so can be used to find other bearings. The sextant, developed in the 17th century, helps navigators to find their latitude and longitude (p. 18) from the altitude of the Sun above the horizon at noon – or the Moon, stars, or planets at night. By taking "sights" of the height of the Sun above the horizon in the daytime, or the heights of well-known stars above the horizon at night, the angle can be calculated and the ship's position fixed.

Rotating fin

Towing ring

Stabilizer fin

Reading in nautical miles

Index mirror

Telescope

Semi-silvered horizon mirror

Handle

Index bar

Reading of heavenly body position above horizon

Angle of arc in degrees and minutes

Fine adjustment screw

CAPTAIN'S LOG
Sailing ships used to measure their speed by heaving a float attached to a line over the side, and measuring how much line ran out in a certain time. Later models used dials which gave a reading in fractions of a nautical mile, depending on how fast the rotating fin spun. When reading the log, the strength and speed of the current of the sea had to be taken into account.

Equator showing degrees of longitude

Model ivory ship

Nautical triangle

WORKING IN CHAINS
As new lands were discovered, they were mapped for future explorers and traders. One of the standard units of measure used to be the chain, 20 m (22 yd) long. Surveyors literally used a chain, like this one, to measure out land. Even using primitive tools like this, remarkably accurate maps were made, as the Roselli map (opposite) shows. The old measure survives today as the length of a cricket pitch.

Compass to set model ship's course

NAUTICAL GLOBE
This "skeleton" globe was made by Charles Hatch in London in 1854. It was used to teach navigation to mariners, to show them what would happen if they set a particular course. Little ivory models of sailing ships can be made to travel along courses, like "nautical triangles", around the globe to indicate the best – and worst – possible sailing routes.

Meridian lines joining North and South Poles

Putting the Earth in perspective

CRYSTAL SPHERE IDEA
Before people discovered that the Sun and stars are at vast distances from the Earth, they thought that stars might be little lights attached to a solid sphere of glassy material around the flat Earth. It would be like a huge glass dome. In those medieval days, people wondered what would happen to a traveller who ventured to the edge of the world and tried to poke his or her head through the dome. However, the Earth is not flat, nor is it covered by a "crystalline sphere". Now it is known that the Sun is so far away that even light, moving at 300,000 km (186,000 miles) per second, takes more than 8 minutes to travel across space to Earth. Even the nearest stars are so far away that light takes several years to travel from them to us.

THE EARTH, THE PLANET on which we live, seems huge to us. The Sun seems to be the most important object in the sky. This is only because humans are so small, and because the Earth orbits so close to the Sun. Our home planet is actually quite small as planets go, and the Sun is really just a very ordinary, unspectacular star. The next nearest star is 7,000 times further away than our Sun. If you could drive non-stop around the Earth at a steady 60 km (37 miles) per hour, the journey would take just under a month. Even at the speed of an airliner, 800 km (500 miles) per hour, it would take 21 years to travel the distance from the Earth to the Sun. To travel from one side of the Solar System to the other, across the diameter of the orbit of Pluto, would take more than 1,600 years at this speed.

Orbital theories

Unlike stars, planets seem to move independently across the sky. The word "planet" is derived from "wanderer" in Greek. If all the stars were on one crystalline sphere, each planet would have to be attached to its own sphere to explain the way planets wander across the pattern of the fixed stars. Throughout history, there have been numerous attempts to explain how this would work, by applying various different forms of mathematical logic. Johannes Kepler, a German astronomer who lived from 1571 to 1630, thought that he could explain the distances to the planets in terms of geometry. Nowadays, our image of the Solar system is based on Newton's theory of gravity, modern telescopic views, and space exploration.

KEPLER'S NESTED ORBITS
Kepler noticed that the way in which regular solids of classical Greek geometry fitted inside each other matched the proportions of the distances to the planets known at that time. One solid has triangular faces, another square faces, and so on. He designed a model to show how these shapes would fit inside the main crystalline sphere of earlier theories. Another sphere, one for each planet, would be between each pair of solids. Kepler's theory failed when he discovered, in 1609, that the planets actually travel round the Sun in ellipses, not circles.

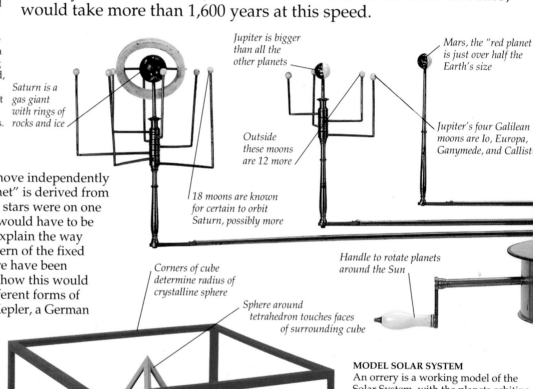

Saturn is a gas giant with rings of rocks and ice

Jupiter is bigger than all the other planets

Mars, the "red planet is just over half the Earth's size

Outside these moons are 12 more

Jupiter's four Galilean moons are Io, Europa, Ganymede, and Callist

18 moons are known for certain to orbit Saturn, possibly more

Handle to rotate planets around the Sun

Corners of cube determine radius of crystalline sphere

Sphere around tetrahedron touches faces of surrounding cube

Dodecahedron has 12 faces of 5-sided pentagons

Octahedron has 8 faces of equilateral triangles

Innermost icosahedron has 20 faces of equilateral triangles

MODEL SOLAR SYSTEM
An orrery is a working model of the Solar System, with the planets orbiting around the Sun, and moons orbiting around the planets. This orrery, made in 1790, shows the four inner planets, Mercury, Venus, Earth, and Mars, and the first three giant planets, Jupiter, Saturn, and Uranus. Neptune was only discovered in 1846, and Pluto not until 1930. The model shows how the planets seem to move along regular paths around the Sun. A powerful new image of the Universe was provided when Isaac Newton (pp. 6 and 28) was able to explain the planetary orbits using his new theory of gravity. The Universe no longer seemed to be composed of mysterious crystal spheres or ruled by capricious gods. For the first time it was seen as an orderly place which could be explained by science.

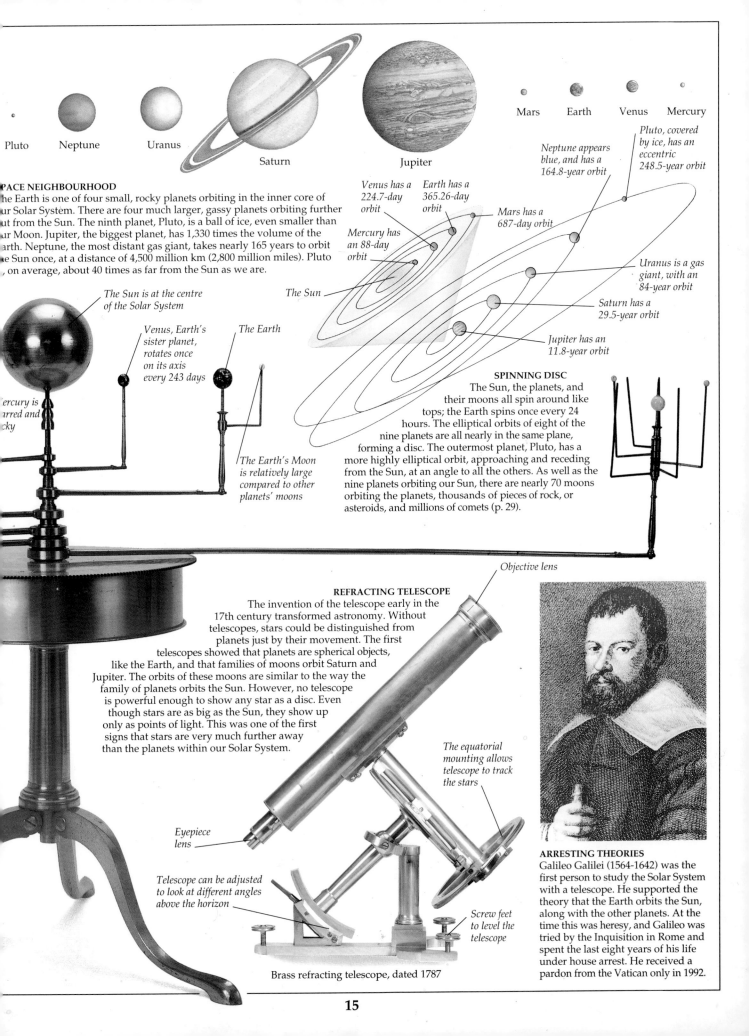

Pluto Neptune Uranus Saturn Jupiter Mars Earth Venus Mercury

PACE NEIGHBOURHOOD

he Earth is one of four small, rocky planets orbiting in the inner core of ur Solar System. There are four much larger, gassy planets orbiting further ut from the Sun. The ninth planet, Pluto, is a ball of ice, even smaller than ur Moon. Jupiter, the biggest planet, has 1,330 times the volume of the arth. Neptune, the most distant gas giant, takes nearly 165 years to orbit e Sun once, at a distance of 4,500 million km (2,800 million miles). Pluto , on average, about 40 times as far from the Sun as we are.

The Sun is at the centre of the Solar System

Venus, Earth's sister planet, rotates once on its axis every 243 days

The Earth

ercury is arred and cky

The Earth's Moon is relatively large compared to other planets' moons

Neptune appears blue, and has a 164.8-year orbit

Pluto, covered by ice, has an eccentric 248.5-year orbit

Venus has a 224.7-day orbit

Earth has a 365.26-day orbit

Mars has a 687-day orbit

Mercury has an 88-day orbit

The Sun

Uranus is a gas giant, with an 84-year orbit

Saturn has a 29.5-year orbit

Jupiter has an 11.8-year orbit

SPINNING DISC

The Sun, the planets, and their moons all spin around like tops; the Earth spins once every 24 hours. The elliptical orbits of eight of the nine planets are all nearly in the same plane, forming a disc. The outermost planet, Pluto, has a more highly elliptical orbit, approaching and receding from the Sun, at an angle to all the others. As well as the nine planets orbiting our Sun, there are nearly 70 moons orbiting the planets, thousands of pieces of rock, or asteroids, and millions of comets (p. 29).

Objective lens

REFRACTING TELESCOPE

The invention of the telescope early in the 17th century transformed astronomy. Without telescopes, stars could be distinguished from planets just by their movement. The first telescopes showed that planets are spherical objects, like the Earth, and that families of moons orbit Saturn and Jupiter. The orbits of these moons are similar to the way the family of planets orbits the Sun. However, no telescope is powerful enough to show any star as a disc. Even though stars are as big as the Sun, they show up only as points of light. This was one of the first signs that stars are very much further away than the planets within our Solar System.

The equatorial mounting allows telescope to track the stars

Eyepiece lens

Telescope can be adjusted to look at different angles above the horizon

Screw feet to level the telescope

Brass refracting telescope, dated 1787

ARRESTING THEORIES

Galileo Galilei (1564-1642) was the first person to study the Solar System with a telescope. He supported the theory that the Earth orbits the Sun, along with the other planets. At the time this was heresy, and Galileo was tried by the Inquisition in Rome and spent the last eight years of his life under house arrest. He received a pardon from the Vatican only in 1992.

Measuring time

TIME IS PERCEIVED AS A LINEAR THING, stretching from the past, through the present, and into the future. Years follow one another in a neat chronological order; Monday is followed by Tuesday, Tuesday by Wednesday, and so on. The passage of hours, one after another, can be seen as a long line, numbered at evenly spaced intervals, like numbers along the edge of a ruler. Our everyday experience of time passing ticks off each number as it passes. The numbers can be chosen to represent centuries, years, months, weeks, days, seconds, or any other regular division of time. Throughout history, people have tried to measure the passage of time, dividing it into smaller and smaller units. For early civilizations dependent on agriculture, it was the passage of the seasons that was important. Daily rhythms of time were important to workers in the fields, and to hunters. Later, more accurate ways of measuring time were needed to count the hours worked by people in factories, and work out how much pay they had earned. Today, scientists measure time in tiny fractions of a second, and "to the minute" accuracy is commonplace in daily life. All of these measurements depend on the fact that under everyday conditions time really does "flow" at a steady, unchanging pace.

TIME IN THE SUN
One of the simplest ways to measure the passage of time is with a sundial. A stick stuck upright in the ground casts a shadow which moves around as the Sun travels across the sky. The shadow is shortest at noon when the Sun is most nearly overhead, and longer in the evening and morning. Both the length of the shadow and its direction give information about the time of day, as well as the season, for these tribesmen in Borneo.

Lid may be removed to turn the clock upside down

Pedestal feet at both ends of the clock

THE SANDS OF TIME
Sand flows at a steady rate through an hourglass, however much – or little – sand is left in the top. The speed the sand flows depends on the size of the hole it flows through. These hourglasses were made in Germany in the 17th century.

Tubes to allow flow of air and water

Incense stick smoulders in dragon

Upper glass container bulb

Lower glass bulb

TIME TRICKLING AWAY
A water clock, or clepsydra, measures an interval of time by the flow of water thruogh a vessel which has a hole in it. This water clock, made in 1670, works using a system of tubes encased within two glass bulbs. When the clock is turned upside down, water from the upper glass bulb runs down into the lower bulb, and air rises up the tubes to replace the water. Constant air pressure ensures a regular flow of water. In some water clocks, the water flows more slowly as the level drops, so that the flow is not constant.

Ball falls to tray as thread burns through

Egyptian
stone tally of
days worked

SLAVES TO TIME

In civilized societies, modern as well as ancient ones, people keep track of time, but time is also used to keep track of people. In many factories, and even some schools, people "clock in" and "clock out" by inserting special cards into automatic registers that keep a record of the workers' – or students' – daily attendance. Although the process has been automated, it is really only a development of keeping a written register. Even when the pyramids were being built, workers had to "clock in" with a foreman, who made a note of which days they worked. The main difference is that today's record shows at exactly which minute the worker arrives, and not just which day.

The fall of each
weight marks time

BURNING TIME

A fire clock can mark the passage of time if it is made of something that burns at a regular, steady rate. A candle, marked with lines corresponding to one hour's burning time, is adequate if it is kept away from draughts. This Chinese fire clock, made in 1800, is even better. An incense stick burns at a very steady rate, and the flame moves past threads which are attached to weights and spaced at regular intervals along the incense stick. As each thread burns, the weights on the end fall into a metal tray – the fire clock equivalent of a long-case clock striking the hours.

*Swinging
pendulum
bob*

*Falling
weight in
velvet bag*

*Long chain
holding
counterweight*

*Silhouette
of pendulum
clock showing
the length
of chain*

SWINGING TIME

The regular swing of a pendulum provided scientists with the first accurate way to measure time, in the 17th century. It was only in the 1650s that Dutch physicist Christiaan Huygens found how to link the regular swing of a pendulum to a mechanism which could drive the hands of a clock in regular ticks. He worked out the theory of pendulums, including the mathematical rule that relates the period of its swing to the length of the pendulum. The first accurate pendulum clock was made to his design in 1657. However, a swinging pendulum swings through a shorter and shorter arc, and eventually stops, because of friction. In this type of pendulum clock, a slowly falling weight is attached to provide energy to keep the pendulum swinging.

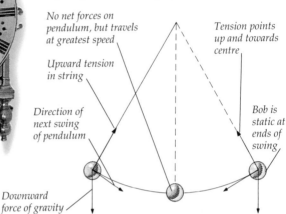

*No net forces on
pendulum, but travels
at greatest speed*

*Tension points
up and towards
centre*

*Upward tension
in string*

*Direction of
next swing
of pendulum*

*Bob is
static at
ends of
swing*

*Downward
force of gravity*

FORCES ON A PENDULUM

Pendulums feel two forces: gravity pulling straight downwards and tension in the string acting upwards at an angle that changes. These can be combined to make one force. At the bottom of the swing, the up and down forces cancel out. The pendulum moves fastest there, but there is no force on it. At each end of its swing, the force is strongest – though the pendulum is not moving at all for a moment, as it reverses.

TIMELY SCIENTIST

Christiaan Huygens' (1629-1695) father was a Dutch diplomat, who gave his son a good education expecting him to follow in his footsteps, but instead he became a scientist. Huygens was supported by his father from 1649 to 1665, when he not only invented the pendulum clock, but began to develop the wave theory of light, and discovered the nature of Saturn's rings.

Like clockwork

BECAUSE THE EARTH ROTATES on its axis once every 24 hours, the Sun seems to move across the sky from east to west. Therefore the time of day depends on where you are on Earth. Noon is when the Sun is highest in the sky; but when it is noon in Tokyo, the Sun has already set in San Francisco, and it is still morning in Bangladesh. There are 360° in a circle and 24 hours in a day, so moving east or west around the Earth by 15° (longitude) changes the time, according to the Sun, by one hour (360 ÷ 24 = 15). The time in New York is 5 hours "behind" London time, because New York lies 75° west of London. Early navigators realized that if they could carry a record of the local time in their home port with them on their travels, they would be able find their longitude on the Earth by measuring the difference between local noon and noon at home. Today, we can find the time anywhere in the world using radio signals, but until the late 18th century, whilst latitude could be calculated by measuring how far the Sun or the Pole Star rose above the horizon, the puzzle of finding longitude was the biggest problem for sailors travelling far from land on long voyages.

TRAVELLING TIME
Before the invention of the electric telegraph enabled fast communication across the country, standard time was carried around the country on trains. Accurate timekeeping was especially important for the postal services, which had to collect and deliver letters punctually. In the UK, clocks in stations where mail trains stopped were set by a mail clock, left, which had been set in London. This was carried in a sealed pouch from town to town. Town hall clocks were set by station clocks, and people set their own clocks and watches by town hall clocks.

Regular upright swing

Pendulum swings further to left

Pendulum swings further to right

GLOBAL COORDINATES
Any location on Earth can be specified by two numbers – the angle north or south of the equator (latitude) and the angle east or west of a chosen north-south line (longitude). Finding longitude at sea, with no fixed geographical features, is very difficult. One way is to compare the local time with time in the home port, but even Isaac Newton (p. 6) thought it would be impossible to make a clock that could be carried around on a rolling ship for months on end. He thought it would stop, or fail to keep an accurate record of the time at home.

Equator

Meridian ring engraved with degrees of latitude

Line of longitude

Early depiction of Australia

Early 17th-century Dutch globe

THE SWINGER SWUNG
Pendulum clocks, like the one invented by Christiaan Huygens (p. 21), will not keep accurate time at sea because the ship is constantly rolling and pitching. As the ship leans to one side, the pendulum swings further "downhill" in that direction. This makes the swing longer, and therefore slower, than it would be if the clock were standing still and upright. At best, a simple pendulum clock on board a ship would keep erratic time. At worst, the long pendulum would probably smash into the side of the clock casing and stop swinging altogether, until another chance roll of the ship set it in motion again.

A CLOCK FOR ALL SEASONS

In 1714 the British government offered a prize of £20,000 to anyone who could find a practical way to determine longitude to half a degree of accuracy on a voyage from Britain to the West Indies. To win the prize, a clock would have to keep time to within 2.8 seconds per day at sea. In the 1730s the English clockmaker John Harrison built a clock he named H1 (below). Tested on a voyage to Lisbon and back in 1736, it passed the test with flying colours. Built to be entirely independent of gravity, its moving parts are counterbalanced and controlled by springs. Harrison had invented a new kind of mechanism to control the clock, and even to compensate for changes in temperature. The clock movement weighs 34 kg (75 lb) and stands 63 cm (2 ft 3 in) high. However, because it was a unique prototype that could not easily be copied, H1 was not accepted as the prizewinner. Harrison continued to work on improved designs, coming up with two new versions. His H3 would probably have won the prize, but it was never entered. By then Harrison had found an even better solution to the longitude problem – the H4 watch.

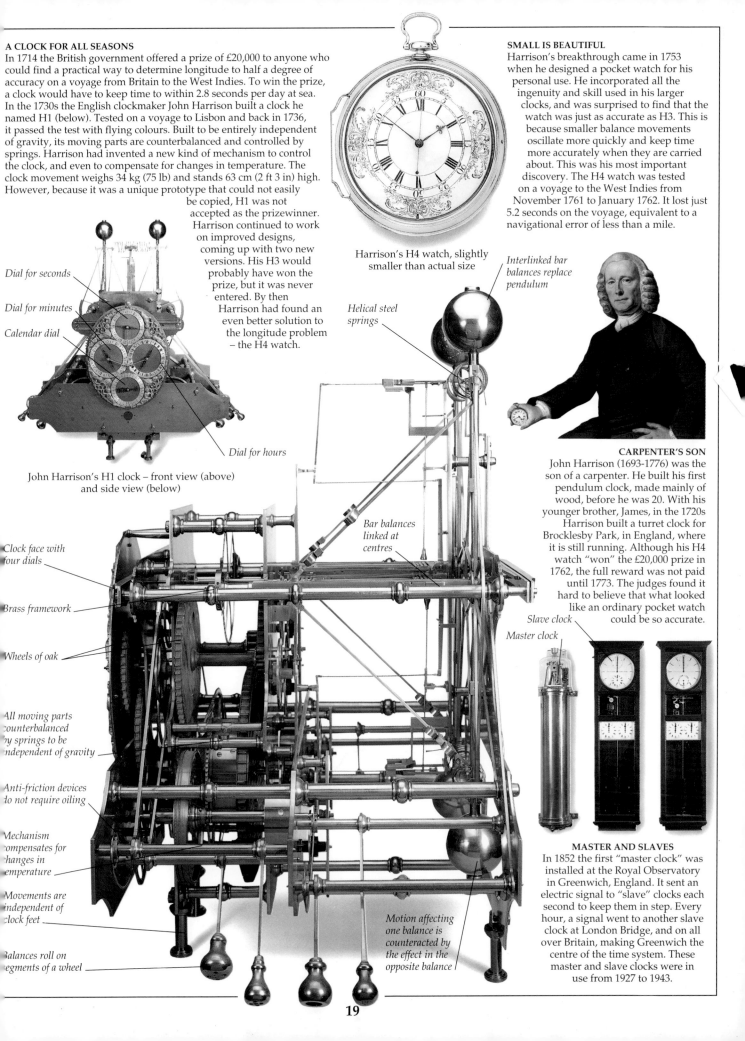

Dial for seconds
Dial for minutes
Calendar dial
Dial for hours

John Harrison's H1 clock – front view (above) and side view (below)

Clock face with four dials

Brass framework

Wheels of oak

All moving parts counterbalanced by springs to be independent of gravity

Anti-friction devices do not require oiling

Mechanism compensates for changes in temperature

Movements are independent of clock feet

Balances roll on segments of a wheel

Helical steel springs

Bar balances linked at centres

Motion affecting one balance is counteracted by the effect in the opposite balance

SMALL IS BEAUTIFUL

Harrison's breakthrough came in 1753 when he designed a pocket watch for his personal use. He incorporated all the ingenuity and skill used in his larger clocks, and was surprised to find that the watch was just as accurate as H3. This is because smaller balance movements oscillate more quickly and keep time more accurately when they are carried about. This was his most important discovery. The H4 watch was tested on a voyage to the West Indies from November 1761 to January 1762. It lost just 5.2 seconds on the voyage, equivalent to a navigational error of less than a mile.

Harrison's H4 watch, slightly smaller than actual size

Interlinked bar balances replace pendulum

CARPENTER'S SON

John Harrison (1693-1776) was the son of a carpenter. He built his first pendulum clock, made mainly of wood, before he was 20. With his younger brother, James, in the 1720s Harrison built a turret clock for Brocklesby Park, in England, where it is still running. Although his H4 watch "won" the £20,000 prize in 1762, the full reward was not paid until 1773. The judges found it hard to believe that what looked like an ordinary pocket watch could be so accurate.

Slave clock
Master clock

MASTER AND SLAVES

In 1852 the first "master clock" was installed at the Royal Observatory in Greenwich, England. It sent an electric signal to "slave" clocks each second to keep them in step. Every hour, a signal went to another slave clock at London Bridge, and on all over Britain, making Greenwich the centre of the time system. These master and slave clocks were in use from 1927 to 1943.

Biotime

MANY LIVING THINGS HAVE an inbuilt sense of time. People have a natural rhythm of sleeping and waking, and manage their day around the 24-hour spin of the Earth (pp. 28-29). This body clock actually runs on a rhythm slightly more than 24 hours long, but people can exercise control over it. Many species, including birds, fish, and land animals, migrate over long distances, in tune with the changing pattern of the seasons, maintaining a regular annual cycle. The natural patterns of biotime also show up in the lifespans of different creatures. In general, smaller creatures live their lives more rapidly, and die sooner. Large animals with a more leisurely lifestyle tend to live longer. Some species, including tortoises and some bears, can slow their normal life processes so much that they can live through the winter months without eating, in a state of hibernation. Within our bodies, different cellular clocks tick at different rates. Liver cells divide every year or two, while the cells that line the stomach divide twice per day. Cells from a human baby, put in a laboratory culture, divide 50 times before they die. Cells from a 40-year-old divide just 40 times, and cells from an 80-year-old divide only 30 times. This suggests that animals are programmed to have a fixed amount of time to live.

BOTANICAL CLOCKMAKER
Carl von Linné (1707-1778), usually known as Linnaeus, was a Swedish botanist and explorer who established the first scientific system of naming plants and animals, organizing them into groups and classes. It is still used today. He also found that flowers of different plants open and close at different times during the day. In 1745 Linnaeus devised a "flower clock" which told the time according to which flowers were open.

TIMELY MIGRATION
Bramblings are songbirds that breed in coniferous and birch woodland in the northern hemispere, from Scandinavia to Japan. They are about 15 cm (6 in) long, and members of the finch family. They migrate south in huge flocks, like this one, in the autumn. Sometimes, millions of bramblings arrive in Europe for the winter. Animals migrate for many reasons – to avoid harsh winter cold or to look for fresh supplies of food. Birds often travel in large flocks, as there is safety in having many pairs of eyes to watch for danger. To set off together, they tell the time by changes in temperature from summer to winter, and the change in the amount of daylight.

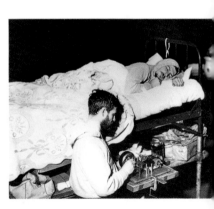

CAVEMAN EXPERIMENT
Unable to see daylight, people naturally follow a circadian rhythm, or body clock, about 25 hours long – close to the 24-hour, 50-minute rhythm of the tides. In experiments at the University of Chicago, volunteers living in an underground room kept a 28-hour "day". They did not adjust well to the new pattern, and experienced symptoms like jet lag.

BUTTERFLIES FLUTTER BY
American Monarch butterflies have a strong inbuilt "clock". In summer, they travel north to Canada, and reproduce. The last generation of summer butterflies migrates south to Florida, Texas, California, and Mexico, travelling more than 3,000 km (1,800 miles). They gather in huge numbers in sheltered spots to hibernate. In spring, they start northward again, but stop to lay eggs and die.

RHYTHMS OF SLEEP
Human rhythms of biotime show up when the electrical activity in the brain of a sleeping person is monitored. There are two distinct phases of sleep: dream, or REM, sleep, and non-REM sleep. There will be four or five stages of REM sleep in an average night. Electrode attached to a volunteer's head can record the changing activity of the brain on an electroencephalograph (EEG). In the deepest REM stage of sleep, the brain waves are slowest and most pronounced. During light sleep, the EEG trace shows more rapid, but less pronounced, flickering. The deepest stage takes up 13 per cent of the sleep of young adults, but older people hardly ever have this amount of deep sleep.

Fresh red pepper

Pepper, 8 days later

Pepper, after 10 days

Rotten pepper, after 15 days

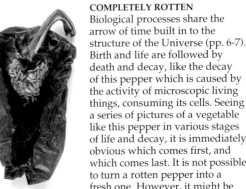

COMPLETELY ROTTEN
Biological processes share the arrow of time built in to the structure of the Universe (pp. 6-7). Birth and life are followed by death and decay, like the decay of this pepper which is caused by the activity of microscopic living things, consuming its cells. Seeing a series of pictures of a vegetable like this pepper in various stages of life and decay, it is immediately obvious which comes first, and which comes last. It is not possible to turn a rotten pepper into a fresh one. However, it might be possible to slow down the ageing, or rotting, processes, with the use of preserving techniques. To make old people – or peppers – young once more, the second law of thermodynamics (pp. 30-31) would be broken, which means that time would have to run backwards.

*ark band shows
yer of coal dust
white
eposit*

*Pale band shows each
gap between each
working shift*

*Wide, pale band
is day of rest*

*Evidence of the
tree rings can
still be seen*

*Ring pattern
indicates it was
a coniferous tree*

*Tree rings indicate
life history of tree*

*Thick rings show
good years for the tree*

HE "SUNDAY STONE"
his rock was formed in a Tyneside coal
nine in England during the 1800s.
white mineral, barium sulphate,
ettled out in a water trough, and
uring working shifts the deposit
vas blackened by coal dust. The
attern of bands indicates a six-
ay working week. The rock
orms a diary of days worked,
o is a permanent record
f human activity.

*Stone is a different
colour where bark
surrounded the trunk*

FROZEN IN TIME
The trunk of a tree is
made from a succession
of growth rings, when
cut horizontally. A tree
grows actively in spring
nd summer, but tends to
lie much more dormant
through the winter. The
ings record its life story.
The number of rings in
he trunk tells us how old
it was, in years, when it
fell. Thick rings show that
the weather that year was
good for growing tees, and
there was plentiful water and
ood to take up in the microscopic
tubes that form the rings; thin rings
pinpoint dry or cold years, so fewer,
smaller, microscopic tubes grew. This
tree trunk is actually a fossil. The rings
were built up in a living tree 200 million
years ago, and have since been turned to stone.

*The wood
of the tree
has been
replaced by the
mineral agate*

Biospace

LIVING THINGS USE SPACE IN DIFFERENT WAYS. Each species has its own ecological niche. This is defined by the type of food it eats, the kind of predators that eat it, the range of temperatures it can tolerate, and its activities. A forest might be "full" of deer, because there is no food for any more deer, but it could still have room for other species, such as rodents, or flowers. The amount and type of space individual members of a species need vary enormously from one species to another. Social insects, like ants, live almost literally on top of one another; some large birds, such as the albatross, range over vast areas of ocean. Plants cannot move around other than by growing, but they also use space in different ways. Some tiny plants cling to cracks in mountain rocks. Others, like mangroves, thrive in coastal swamps. If two species are in competition, one of them will have to find an alternative niche, or it is likely to become extinct. Wherever there is food and water on Earth, there is life. One way or another, life expands to fill the space available.

ECOLOGICAL ECONOMIST
Thomas Malthus (1766-1834), an English economist, went to Cambridge University and was ordained in 1788 – but became a Professor of History and Political Economy in 1805. While working as a curate, in 1798, he published *An Essay on the Principle of Population*, which argued the human population would eventually grow so much that it would use up the world's resources.

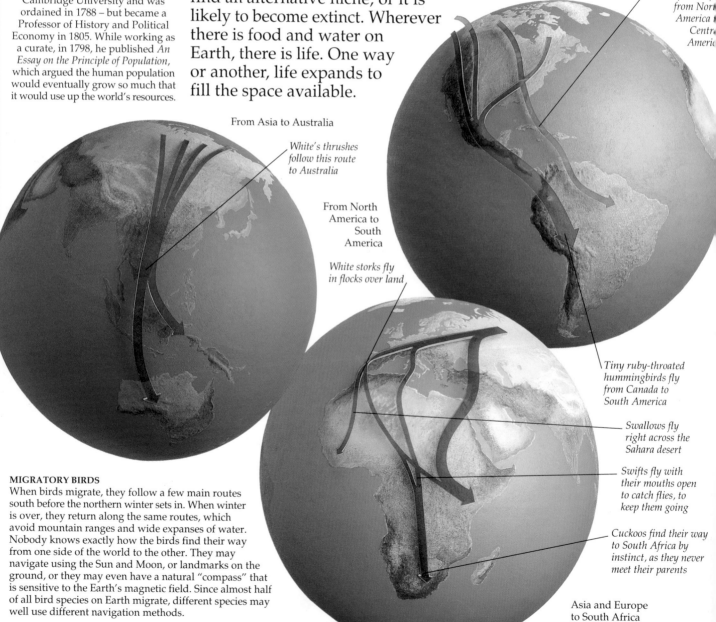

From Asia to Australia

White's thrushes follow this route to Australia

From North America to South America

White storks fly in flocks over land

Golden plovers fly from North America to Central America

Tiny ruby-throated hummingbirds fly from Canada to South America

Swallows fly right across the Sahara desert

Swifts fly with their mouths open to catch flies, to keep them going

Cuckoos find their way to South Africa by instinct, as they never meet their parents

Asia and Europe to South Africa

MIGRATORY BIRDS
When birds migrate, they follow a few main routes south before the northern winter sets in. When winter is over, they return along the same routes, which avoid mountain ranges and wide expanses of water. Nobody knows exactly how the birds find their way from one side of the world to the other. They may navigate using the Sun and Moon, or landmarks on the ground, or they may even have a natural "compass" that is sensitive to the Earth's magnetic field. Since almost half of all bird species on Earth migrate, different species may well use different navigation methods.

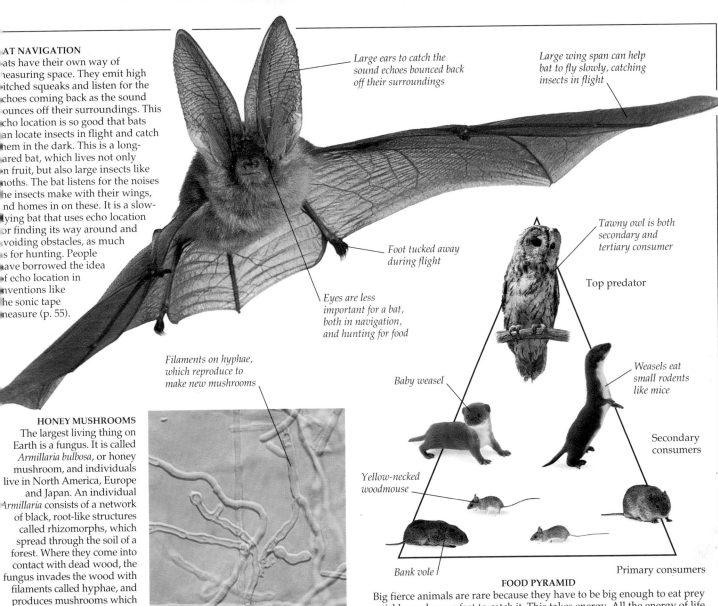

BAT NAVIGATION

Bats have their own way of measuring space. They emit high pitched squeaks and listen for the echoes coming back as the sound bounces off their surroundings. This echo location is so good that bats can locate insects in flight and catch them in the dark. This is a long-eared bat, which lives not only on fruit, but also large insects like moths. The bat listens for the noises the insects make with their wings, and homes in on these. It is a slow-flying bat that uses echo location for finding its way around and avoiding obstacles, as much as for hunting. People have borrowed the idea of echo location in inventions like the sonic tape measure (p. 55).

Large ears to catch the sound echoes bounced back off their surroundings

Large wing span can help bat to fly slowly, catching insects in flight

Foot tucked away during flight

Eyes are less important for a bat, both in navigation, and hunting for food

Tawny owl is both secondary and tertiary consumer

Top predator

Weasels eat small rodents like mice

Baby weasel

Secondary consumers

Yellow-necked woodmouse

Bank vole

Primary consumers

FOOD PYRAMID

Big fierce animals are rare because they have to be big enough to eat prey quickly, and move fast to catch it. This takes energy. All the energy of life on Earth comes from the Sun, and is trapped by plants. The stored energy passes up the food chain, but some is lost at each stage. The "chain" is really a pyramid. At the bottom of the pyramid are primary consumers, eating plant material. The next layer, which ecologists call a trophic level, will be a secondary consumer, which feeds on the herbivores. At the top of this pyramid, the owl is a secondary and tertiary consumer, as it eats both voles and weasels. The big hunters are much rarer than the herbivores.

HONEY MUSHROOMS

The largest living thing on Earth is a fungus. It is called *Armillaria bulbosa*, or honey mushroom, and individuals live in North America, Europe and Japan. An individual *Armillaria* consists of a network of black, root-like structures called rhizomorphs, which spread through the soil of a forest. Where they come into contact with dead wood, the fungus invades the wood with filaments called hyphae, and produces mushrooms which grow on the decaying wood and produce spores. The fungus digests the dead wood from inside. The biggest individual known has a network of rhizomorphs over 15 hectares (37 acres), and is estimated to weigh 11 tonnes, not counting the mushrooms.

Filaments on hyphae, which reproduce to make new mushrooms

Microscopic view of *Armillaria*

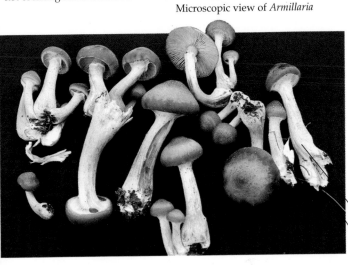

"Fruiting bodies" of *Armillaria* (mushrooms)

Edible honey mushroom

Rhizomorphs spread through soil, under the visible mushrooms

NICHE DESTRUCTION

When rainforest is cut down, it isn't only trees that are destroyed. The forest is a whole ecosystem, with many kinds of plant and animal living together. A study of one plot 25 m by 25 m (27 yd x 27 yd) in West Africa found 190 species of plant, all using the same space in different ways. Rainforest animals include pigs, monkeys, many birds, deer, reptiles, millipedes 30 cm (1 ft) long, beetles the size of mice, and countless other invertebrates.

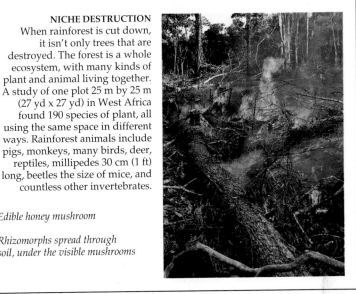

Timescales of history

How long is the history of time? In the 17th century Archbishop James Ussher calculated from clues that he found in the Bible that the creation of the Universe had occurred in 4004 BC. If he was right, by now the history of the Earth has lasted only about 6,000 years, but evidence from astronomy, biology, and geology shows that the Archbishop was wrong. The Earth has been around for nearly five billion years, and the whole Universe has been around for three times as long. Even so, the evidence suggests that there really was a definite beginning to the Universe (pp. 60-61) – which means there is an "edge" of time. To put this in perspective, if all of the Earth's history could be represented by the old English yard – the distance from the King's nose to the tip of his outstretched finger – then a single stroke of a nail file across the King's middle finger would remove all of human history. Biologist Richard Dawkins has put this in another perspective. Five billion years of evolution on Earth have produced people out of primordial soup. If a single human pace represented a thousand years of human history, then the length of the "road" since life on Earth began to the present day would be equivalent to walking all the way from London to Beijing.

STAR OF BETHLEHEM
The Bible tells us that a bright star appeared in the sky to mark the birth of Jesus Christ. At that time, Chinese astronomers chronicled unusual sights and changes in the sky. Interpretation of those records tells us that there might really have been a "new" star visible for a few weeks at just about the right time. However, stars lie so far away from us across space that light from them may take hundreds, or thousands, of years to reach Earth. The star that was visible on Earth at the time Jesus was born must have flared up centuries before.

Exploding star

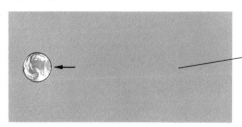

Light reaches Earth after death of star

HAS-BEEN STARS
Some stars flare up brightly just before they die. Such a bright "new" star may be a supernova, or nova. The light from the star has to travel across space before it reaches the Earth. Light travels at 300,000 km (186,000 miles) per second, but by the time the light reaches telescopes on Earth the star may have been dead for hundreds of years. Looking at the stars is like looking back into the past. Because astronomers know how far away stars are, and how long it takes light to travel such distances, they know the Universe is much more than 6,000 years old.

LIVING ROCK
Biology and geology both suggest that the Earth is very old. It takes a very long time – from thousands to millions of years – for sediments laid down in the sea to form thick layers of rock. It must have taken millions of years for the primitive life forms found as fossils in ancient rocks to evolve into the kinds of living things seen today. In 6,000 years, or even in 6 million years, there would not have been time for all the forms of fossil organisms that we now know to have lived their lives, one after the other. These ammonites fossilized in limestone are from the early Jurassic period, around 200 million years ago.

Dead star

Spiral shape of fossilized ammonite

A SEEKER OF ORIGINS
Charles Darwin (1809-1882) travelled round the world from 1831 to 1836 as the naturalist on board the survey ship HMS *Beagle*. The variety of living things that he saw on the voyage helped him to develop the theory of evolution by "natural selection". He saw that species had adapted to different ways of life, even though they had a common ancestor. He realized that new species could arise as a result of the build-up of small changes from one generation to the next. He also realized that this must be a very slow process. For example, a creature the size of a mouse could evolve into a creature the size of an elephant in 60,000 years, from small, almost unnoticeable, changes at each new generation.

STAR AND WEATHER MAN

Heinrich Olbers (1758-1840) was a German doctor and astronomer. His main astronomical passion was studying comets, and he was also interested in the effect of the Moon on the weather. He is mainly remembered today because in 1823 he drew attention to a puzzle that is now known as "Olbers' Paradox". Olbers was not the first person to think up the puzzle, but he did publicize it. The puzzle is: why is the sky dark at night? Olbers realized that if the Universe is infinite and full of stars, then everywhere we look, at every single point in the sky, we should see a star.

THE SKY AT NIGHT

The real sky at night shows the stars scattered across a dark background. There are no stars visible "through the gaps" between stars that are known. Astronomers now know that stars are grouped together in galaxies, and these are distributed across space. Olbers' Paradox applies just as forcefully to galaxies as to stars; however, the spaces between stars and galaxies are huge. It takes starlight a long time to travel across space, and although there are many millions of galaxies in space, there has not been enough time since the Universe began for the light from all the stars to "fill up" space and make the sky blaze with light all the time. The "edge" of the Universe is an edge in time, not in space, meaning the Universe had a beginning.

STARRY NIGHT

If the Universe has always existed and extends forever, it would be full of stars, and everywhere we look we should see a star. Some of the stars will be nearby, and some far away. Every "line of sight" out into space must end on a star, making the sky blaze even at night. At first, astronomers thought that the sky must be dark because there is an "edge" to the Universe, with no stars to be seen beyond the edge.

Night sky appears light from the blaze of the stars

View within "infinite" wood

Nothing can be seen beyond the trees

Gaps can be seen through the trees from any position in a small wood

Wood with an "edge" to it

SEEING THE WOOD FOR THE TREES

Olbers' Paradox can be compared to the saying of "not being able to see the wood for the trees". In the middle of a large, dense forest, every "line of sight" is blocked by a tree trunk. There are only gaps where no trees are visible, near the edge of the wood. If the wood went on forever, no gaps would ever be visible at all. As gaps can be seen between the stars, astronomers of the 19th century made the mistake of thinking that there must be no stars, or galaxies, beyond a certain distance from the Earth in the Universe.

Fields beyond can be seen through the trees

View within small wood

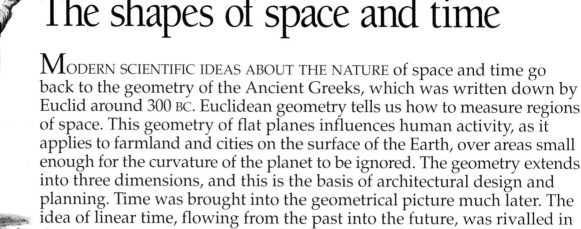

The shapes of space and time

MODERN SCIENTIFIC IDEAS ABOUT THE NATURE of space and time go back to the geometry of the Ancient Greeks, which was written down by Euclid around 300 BC. Euclidean geometry tells us how to measure regions of space. This geometry of flat planes influences human activity, as it applies to farmland and cities on the surface of the Earth, over areas small enough for the curvature of the planet to be ignored. The geometry extends into three dimensions, and this is the basis of architectural design and planning. Time was brought into the geometrical picture much later. The idea of linear time, flowing from the past into the future, was rivalled in the Ancient Greek culture, and also in the Chinese, Hindu, and Aztec cultures, by the idea of cyclic time. This says that history repeats itself, with an ongoing succession of golden ages and dark ages.

INFLUENTIAL MATHEMATICIAN
Euclid lived between around 330 BC and 260 BC, and was not a great mathematician himself, but he wrote down everything the Greeks knew about geometry in a series of 13 books. Translated into Arabic, they survived the Dark Ages. They were then translated into Latin and other languages. Euclid's geometry was the basis of all mathematics for 2,000 years. It still provides the best insight into the marriage of space and time.

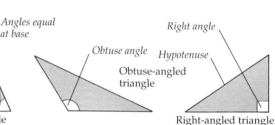

Angles equal at base

Obtuse angle

Right angle

Hypotenuse

Obtuse-angled triangle

Isosceles triangle

Right-angled triangle

TRIANGULAR GEOMETRY
Euclid's "common sense" ideas apply to flat surfaces, and the geometry of triangles is a key to understanding the shapes of space and time. The angles of a triangle always add up to exactly 180°. Isosceles triangles have two angles the same as each other, and two sides the same length. In calculations involving isosceles triangles, a line from the top, or apex, to the middle of the base divides it into two mirror-image right-angled triangles. If one of the angles in a triangle is bigger than 90°, the triangle is said to be obtuse. Right-angled triangles are the most important, where Pythagoras' theorem (p. 10) applies. To find the shortest distance between two points, draw a right-angled triangle connecting them. The separation is the hypotenuse, which can be found using Pythagoras' theorem (p. 14).

North burial barrow under footpath

Aubrey stone

Main avenue leads off to the east

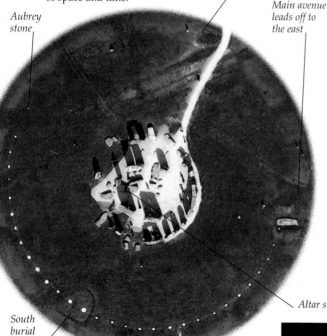

South burial barrow

Altar stone

TRIANGULATION
Euclidean geometry is used by surveyors to measure distances to objects that are far away. By measuring the distance between the two flags, and the angles from each end of this baseline to the house over the river, the distance to the house can be found without crossing the river. Triangulation can even be used to measure distances to the Moon and Mars.

Measured angle

Measured line

Measured angle

STONEHENGE
The oldest working observatory in the world was built on Salisbury Plain in England between 2600 BC and 1700 BC. There seem to have been three main periods of building. It was completed over 1,000 years before Euclid was born. Stonehenge shows our ancestors' interest in the nature of time. Nobody knows why it was built, but astronomer Sir Fred Hoyle (p. 39) has shown that by moving marker stones around the outer ring Stonehenge can be used to predict eclipses. The large, inner stones are aligned geometrically so that on midsummer day the Sun rises exactly on the line of the main avenue.

THE GEOMETRY OF CITIES
City planners who lay out streets and avenues on a grid, so that they cross each other at right angles, are using Euclidean geometry. This view of Los Angeles at night clearly shows a classic gridded city, the street lights defining the lines of the avenues and intersections. Scientists imagine space to be filled with a grid of lines like this, and describe the shapes of space and time by the way the imaginary grid lines are bent, making them "non-Euclidean" (pp. 40-41).

RIOTS OVER LOST TIME

Because there is not a whole number of days in a year, just over 365¼, calendars gradually get out of step with the seasons. In 1752 the English calendar was so far out of step that 11 days had to be skipped, so the day following September 3 became September 14! Many ordinary people thought their lives had been shortened and days of paid work lost, so there were riots as they demanded the return of their "lost" 11 days. Hogarth, a cartoonist, drew satirical scenes at the time, left. Since then, the calendar has been kept in step by omitting leap years in all centenary years except those that divide by 400. So 1900 was not a leap year, but 2000 will be.

Mask of Aztec god, Tonatiuh

Days of the Aztec year

LUNAR ECLIPSE

Before eclipses were understood, they seemed to be the work of capricious gods. By learning how to predict them, people brought order into the Universe. Even before it was understood how eclipses happened, repeating patterns were found that made it possible to predict them, which is an early example of human conquest of time.

Time-lapse photograph of partial lunar eclipse

CYCLIC AZTEC CALENDAR

Over 500 years ago, the Aztec civilization of Central America developed a calendar far more accurate than the European calendar of that time. This calendar stone has all the signs of the days in the Aztec year in a ring. The outer ring of serpents represents the material universe, the next ring the Sun's rays. The calendar is the second ring from the inside, with signs representing days of the year. In the centre is a mask of one of their gods, Tonatiuh, surrounded by symbols that depict the earthquake which the Aztecs believed would end the world.

THE GEOMETRY OF BUILDING

Geometrical shapes were very important in the Middle Ages. People seemed to be at the mercy of nature in their everyday lives, but they could see humankind imposing order on nature in the form of great buildings like this church of St Serge, built at Angers, in France, early in the 13th century. Although many architectural advances were the result of trial and error, buildings like this show how two-dimensional geometry extends into three dimensions. Making roofs to cover spaces as large as a church, which would be much bigger than most domestic houses, presented a problem to architects and builders who usually only worked in wood and stone. Roof shapes like these vaults evolved, which needed a complex application of Euclidean geometry in three dimensions, showing an early example of human conquest of space.

Domed rib vault

Four ribs of stone span the aisle

Corbel juts out from the side wall to support the rib

Buttress gives extra support to the side wall

Central stone where all ribs meet

Gable end wall uses triangular geometry

Stone fills in between ribs

Central stone holds arch together

Transverse arch

Columns on both sides of nave support vaulted roof

Internal space is divided into separate smaller spaces

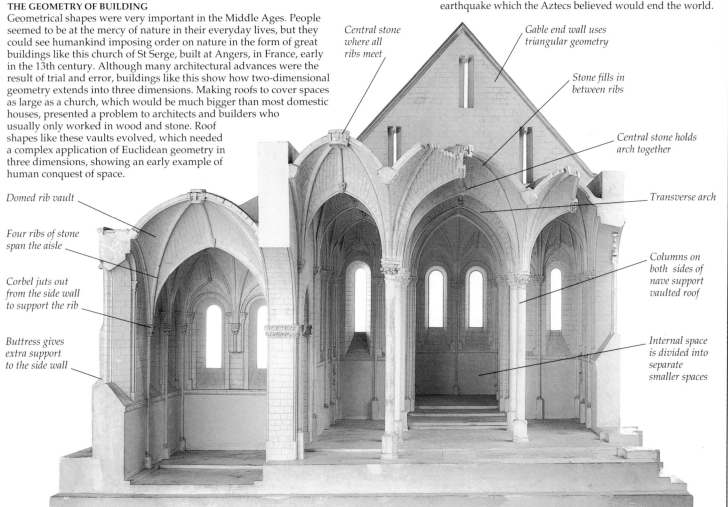

Putting the Universe in order

Isaac Newton discovered the rules on which the Universe runs. Before his time, nobody could explain how the stars and planets moved through the heavens. They seemed to be operated by the whims of capricious gods, or to be revolving on the surfaces of invisible crystal spheres. In his grand theory of gravity, Newton provided one scientific rule that explained the movements of all the heavenly bodies, as well as the behaviour of falling objects on Earth. He also formulated the three laws of motion, describing how objects move when they are pushed or pulled by forces. All his theories and laws were expressed mathematically, and they effectively put the Universe in order. Using Newton's laws of motion and gravity, astronomers were able to predict the positions of planets – and comets – in years to come, and they could calculate where they were in years gone by. Newton believed that all motion took place against a background of "absolute time", and that objects moved through a fixed "absolute space".

A GREAT SCIENTIST
After setting the Universe in order, Newton (pp. 6-7) became Master of the Royal Mint, and he helped reform the British economy. He was also President of the Royal Society in London for more than 20 years.

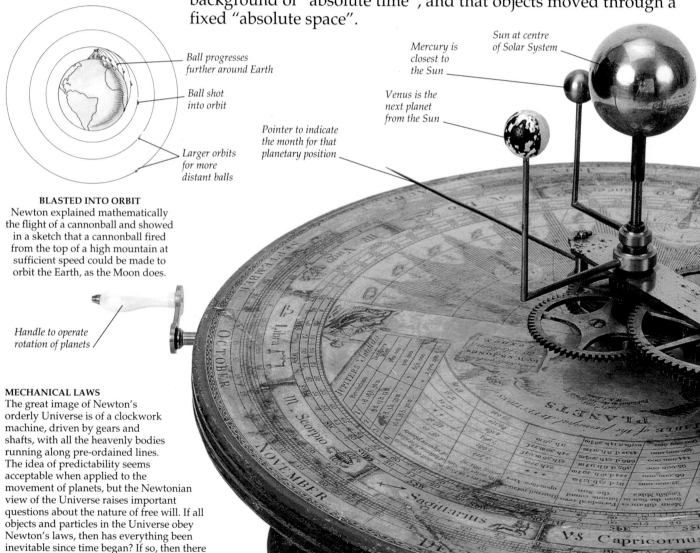

Ball progresses further around Earth

Ball shot into orbit

Larger orbits for more distant balls

Sun at centre of Solar System

Mercury is closest to the Sun

Venus is the next planet from the Sun

Pointer to indicate the month for that planetary position

BLASTED INTO ORBIT
Newton explained mathematically the flight of a cannonball and showed in a sketch that a cannonball fired from the top of a high mountain at sufficient speed could be made to orbit the Earth, as the Moon does.

Handle to operate rotation of planets

MECHANICAL LAWS
The great image of Newton's orderly Universe is of a clockwork machine, driven by gears and shafts, with all the heavenly bodies running along pre-ordained lines. The idea of predictability seems acceptable when applied to the movement of planets, but the Newtonian view of the Universe raises important questions about the nature of free will. If all objects and particles in the Universe obey Newton's laws, then has everything been inevitable since time began? If so, then there is no such thing as free choice, and a physicist or a mathematician with enough information could predict that you would now be reading this paragraph about Newton's laws!

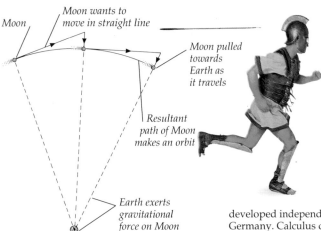

Moon

Moon wants to move in straight line

Moon pulled towards Earth as it travels

Resultant path of Moon makes an orbit

Earth exerts gravitational force on Moon

HELD IN GRAVITY'S GRIP

The Moon is a remote and mysterious object. Newton's greatest achievement was to realize that it obeys the same laws as an object on the surface of the Earth, like a cannonball. The Moon's path "falls" like a cannonball, but it falls very gently because it is so far from the centre of the Earth. At the surface of the Earth, a cannonball would fall 4.9 m (16 ft) in the first second of its fall, since the acceleration due to gravity is 9.8 m/s^2 (32 ft/s^2). The Moon is 60 times further away from the centre of the Earth, so using Newton's inverse square law of gravity it "falls" by a little more than 1.3 mm (0.05 in) in one second. Travelling at the speed of the Moon, at the Moon's distance, a sideways nudge of this size every second is enough to make it travel in a closed orbit around the Earth, instead of heading off into space.

CATCHING THE RUNNER

To calculate the Moon's orbit, Newton needed to add up lots of tiny changes accurately. Like the arrow chasing the runner (pp. 8-9), if the Moon was stationary at the beginning of every second, adding up all of its tiny "falls" does not give the right answer. Newton invented the technique now called calculus to do the job. Calculus was also developed independently by Wilhelm Leibniz of Germany. Calculus can add up an infinite number of infinitely small changes correctly – and it also says the arrow will catch up with the runner!

Planet

Line joining planet and the Sun

Closer to the Sun, the planet moves faster

SWEEPING ORBITS

The nearer a planet is to the Sun, the more strongly the Sun's gravity tugs on it, and the faster it moves in its orbit. An imaginary line joining the planet and the Sun always sweeps out the same area in the same time. This clue helped Newton discover the correct law of gravity, which is now known as the inverse square law, and can be used to calculate why the Moon orbits the Earth, rather than shooting off in a straight line into space.

Further away, orbiting object moves slower

The Sun

In the same time, all four areas swept out are equal

Earth spins once on its axis every 24 hours as it revolves around the Sun

The Moon orbits the Earth, held by its gravity

Halley's comet, seen in 1910, in a photograph tinted in later years

Halley's comet appeared in 1066, commemorated in the Bayeux Tapestry

Disc showing the calendar and zodiac

HALLEY'S COMET

The astronomer Edmond Halley encouraged Newton to publish his discoveries. Halley realized that all heavenly bodies must obey Newton's law of gravity. He decided that four "comets" which had been seen in 1456, 1531, 1607, and 1682 were really different visits of the same object, and predicted the comet would return in 1759. It did, 17 years after his death, and was named after him. This was a triumphant confirmation that Newton's law of gravity really is universal.

The ultimate law of nature

THE MOST IMPORTANT LAW OF SCIENCE is that things wear out. Living things grow old and die, cars rust, and buildings fall down if they are not maintained. Another way of saying this is that the amount of disorder in the Universe increases. Scientists measure disorder as a quantity called entropy, and entropy always increases. This is the second law of thermodynamics. The first law simply says that energy cannot be created or destroyed, just changed from one form to another. The increase of entropy in the Universe shows which way time is flowing: more disorder corresponds to a later time.

When anything is manufactured, it looks as if the second law is being broken, but the new orderliness is always more than made up for by new mess elsewhere, in mining raw materials and generating the energy to produce the object, and so on.

RUST TO DUST
The passage of time shows up in nature in the form of decay. Rusty cars are never seen becoming shiny and rust free, just as a pile of old bricks are never seen assembling themselves into a house, without human help. The opposite processes of cars rusting and buildings falling down are common. If you are shown a picture of a shiny new car next to one of the same car rusty and derelict, it is obvious which was taken first. Time is fundamentally built in to the laws of nature.

AS COLD AS ICE
Heat always flows from hot to cold, never the other way. Ice cubes melt in warm water, and the water cools. Ice will never give heat to water, so that the ice cools and the water boils. After the ice melts, the energy within the glass has spread out more evenly.

BREAKING DOWN THE ORDERED SYSTEM
Things wear out because they are constantly being shaken up by the environment. Houses are buffeted by wind and rain; ice cubes in a glass of water are buffeted by molecules of water. This buffeting of an orderly system can be represented by a loose-fitting jigsaw puzzle being shaken up. It would be astonishing if the picture could be reconstructed simply by jiggling the pieces at random in a box.

1 ORDERED UNIVERSE
Entropy measures the amount of disorder in the Universe, and the second law of thermodynamics states that disorder increases as time passes. This increase in entropy is the same as a loss of information. There is a great deal of information in the ordered picture on a completed jigsaw puzzle.

2 CRUMBLING UNIVERSE
If a loose-fitting jigsaw puzzle is shaken up, the pieces start to come apart. Constant jiggling breaks the picture into a jumble. A portrait of a human face is still recognizable, but it is much harder to tell the identity of the person. Information is being lost. Entropy is increasing. Time is passing.

BRIGHT SOURCE
The Sun is hot and space is cold, so the Sun radiates energy out into space, from the hotter place to the cooler place. The distinction between the hot Sun and cold space is a kind of order. Eventually, when the Sun has radiated all its energy and cooled down, there will be less order in the Universe, and more entropy. An increase in entropy means useful energy is being used up and turned into a useless form.

ORGANIZING ENTROPY
Life on Earth "feeds" off the entropy from the Sun. When plants use sunlight to provide energy for growth, they make some order in the Universe, and entropy decreases. The decrease in entropy on Earth, however, is just a tiny bubble in a vast sea of increasing entropy. The rate at which the Sun is "wearing out" more than makes up for the tiny amount of order represented by life on Earth.

ONE-WAY PATH OF TIME
The direction in which the arrow of time points seems to be built into the activity of the world – and it always points in the direction of mess and confusion. When a glass falls from a table and breaks, the orderly arrangement of the glass and its contents is broken into disorder. The whole process does not have to be seen in order for us to know that the broken glass on the floor comes later in time than the intact glass on the table. Broken glasses never assemble themselves and jump up in the air to land on the table. More disorder always corresponds to later times.

Ordered system at beginning of time period

Contents spill from glass, increasing disorder

Contents no longer distinct from glass

Contents leave glass

LORD OF THERMODYNAMICS
Thermodynamics shows how energy can be made to do work. During the 19th century this was an important factor in practical science, when William Thomson, who later became Lord Kelvin (1824-1907), worked out the laws of thermodynamics, including the second law. He was also a pioneer in electrical engineering, and designed the first successful transatlantic telegraph cable.

Glass smashes irreparably

Picture no longer discernible

3 CHAOS PREVAILS
Eventually the picture is a complete jumble. Entropy is as big as possible, and there is no information left. Time has stopped for the picture. In the wider Universe, when all the stars have given up all their heat, and the temperature is the same everywhere, it will be in a state of maximum entropy at the end of time.

PUTTING ORDER INTO CHAOS
The jigsaw puzzle can be reconstructed, putting the information back in and taking the entropy out, piece by piece. The order made in the jigsaw is always less than disorder elsewhere in the Universe which results from keeping the person who is doing the puzzle alive. The energy to maintain life comes from food eaten, which comes in turn from sunlight because the Sun is wearing out.

Massive input of energy to recreate order

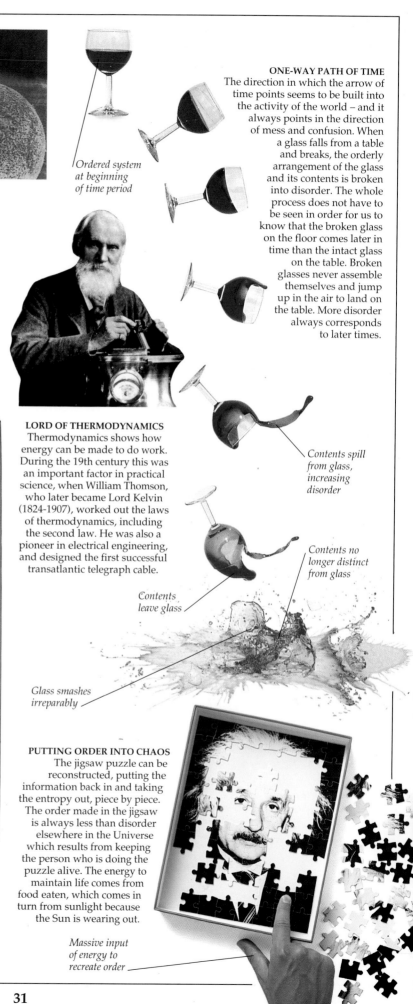

The ultimate speed limit

IN THE EARLY 19TH CENTURY Michael Faraday showed that electric and magnetic effects could be described in terms of "lines of force" reaching out from charges and magnets. In the 1860s James Clerk Maxwell then developed a set of equations describing how ripples in the lines of force would travel across space. His theory linked electricity and magnetism directly, in the relationship now called electromagnetism. He showed how moving magnets create electric currents, while moving charges create magnetic fields. His equations automatically included a fixed number which corresponds to the speed with which electromagnetic ripples move. The number turned out to be the speed of light. Without trying, Maxwell had shown that light is a form of electromagnetic wave. The theory implied that there should be other forms of electromagnetic wave, all travelling at the same speed. This was confirmed with the discovery of radio waves in the 1880s. Now it is known that there is a whole range of electromagnetic waves, called the electromagnetic spectrum, including infrared, ultraviolet, X-rays, and microwaves.

THE MOONS OF JUPITER
Above are two of Jupiter's moons: Europa, visible against the dark background of space, and Io, silhouetted against the planet itself. Although 17th-century astronomers did not have telescopes good enough to show the moons in such detail, they were still able to use studies of their movement, particularly Io, to measure the speed of light. The measurement was made before Isaac Newton published his theory of gravity (p. 28).

SPEEDING LIGHT
Maxwell's electromagnetic theory says that the speed of light is always the same, c. For example, if a car is being driven at 90 km per hour (56 mph), the light from its headlamps is not moving at c plus 90, but simply at c. If the driver measures the speed of the light from the headlamps, he will get the same "answer", c. Tests like the Michelson-Morley experiment prove that Maxwell was right.

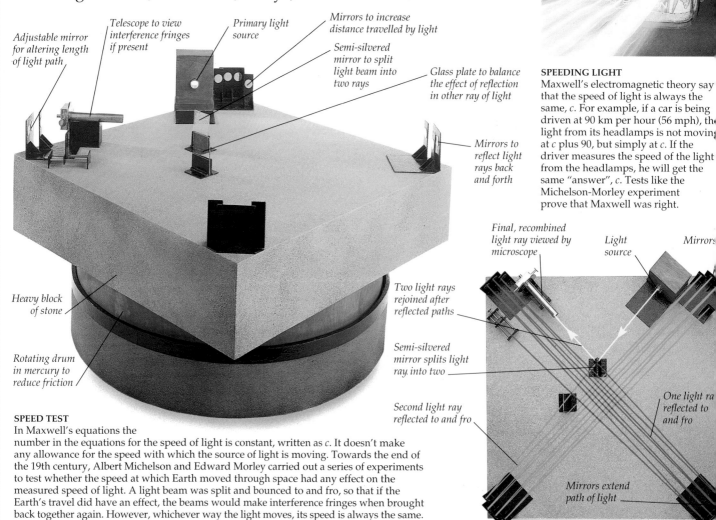

Adjustable mirror for altering length of light path

Telescope to view interference fringes if present

Primary light source

Mirrors to increase distance travelled by light

Semi-silvered mirror to split light beam into two rays

Glass plate to balance the effect of reflection in other ray of light

Mirrors to reflect light rays back and forth

Heavy block of stone

Rotating drum in mercury to reduce friction

Final, recombined light ray viewed by microscope

Light source

Mirrors

Two light rays rejoined after reflected paths

Semi-silvered mirror splits light ray into two

Second light ray reflected to and fro

One light ray reflected to and fro

Mirrors extend path of light

SPEED TEST
In Maxwell's equations the number in the equations for the speed of light is constant, written as c. It doesn't make any allowance for the speed with which the source of light is moving. Towards the end of the 19th century, Albert Michelson and Edward Morley carried out a series of experiments to test whether the speed at which Earth moved through space had any effect on the measured speed of light. A light beam was split and bounced to and fro, so that if the Earth's travel did have an effect, the beams would make interference fringes when brought back together again. However, whichever way the light moves, its speed is always the same.

DANISH ASTRONOMER
The Dane Ole Rømer (1644-1710)
worked at the Paris Observatory in
the 1670s. He was able to stun his
colleagues by predicting, correctly,
that an eclipse of Io by Jupiter, due
on November 9 1679, would be
10 minutes late. He explained this
as due to the time it took light from
Io to reach Earth, and calculated its
speed as 225,000 km (139,810 miles)
per second. Rømer became famous
as a result of this measurement, and
was appointed Astronomer Royal to
King Christian V in Denmark.

CALCULATING "C"
When the Earth is on the opposite
side of the Sun from Jupiter, light
reflected from Jupiter's moons
has extra distance to travel
before reaching the Earth. In
1672 Giovanni Cassini made the
first comparatively accurate
measurement of the distance
from the Earth to the Sun.
Rømer used this to calculate
the speed of light, c. Using
the modern, more accurate
value of 149,597,910 km
(92,960,140 miles) from the
Sun to the Earth, Rømer's
method gives 298,000 km
(185,180 miles) per
second as the speed
of light, which is
remarkably close to the actual value of
299,792 km (186,000 miles) per second. It
was another two centuries before the next
nearest accurate measurement was made.

Jupiter

*Jupiter's moon, Io,
about to be eclipsed*

*Light from Io has
less far to travel,
when on the same
side of the Sun
as the Earth*

*The Earth on
same side of
the Sun as
Jupiter*

*The Earth on opposite
side of the Sun
from Jupiter*

*Light from Io has
further to travel*

The Sun

MAGNETIC PERSONALITY
Aged 15, Scotsman James Clerk Maxwell
(1831-1879) invented an ingenious
method of drawing a perfect oval using
two drawing pins and a loop of thread.
He began research on colour vision,
and made the first colour
photograph, of a tartan. He also
explained the behaviour of gases
in terms of molecules, and deduced
the nature of Saturn's rings as being
composed of solids. After extensive
research into electromagnetism,
in 1874 he founded the pioneering
Cavendish Laboratory in Cambridge,
England, influential for many generations.

RADIO RECEIVER
Light is just one form of
electromagnetic radiation.
Heinrich Hertz discovered radio
waves in 1888, supporting Maxwell's
prediction that light represents only a
small proportion of all electromagnetic
waves. Radio waves have longer
wavelengths, and can be used to send
signals. This 1925 radio has a large
antenna made from several loops
of wire which gather in the radio
waves as they pass. All
electromagnetic waves
travel through space at the
same speed, c, 299,792 km
(186,000 miles) per second.

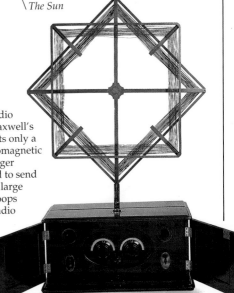

The marriage of space and time

IN 1905 EINSTEIN'S SPECIAL THEORY of relativity united space and time in one mathematical description. In 1907 Hermann Minkowski realized that this was equivalent to treating time as a fourth dimension, in some sense at right angles to each of the three space dimensions. He described relativity theory in terms of geometry. The special theory makes the curious prediction that moving rulers shrink and moving clocks run slow. Minkowski's idea explains this. Anything that exists in space and time has a four-dimensional "length", called extension. This is calculated using the four-dimensional version of Pythagoras' theorem. Extension is a fixed property. The amount of extension that shows up as three-dimensional length, and the amount that shows up as a time interval, depends on the perspective of a moving observer. The two always balance out.

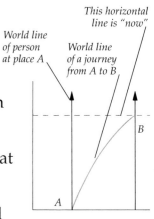

This horizontal line is "now"

World line of person at place A

World line of a journey from A to B

B

A

WORLD LINES
Minkowski showed how to represent spacetime on paper. "Upwards" is the passage of time, and "sideways" represents movement through space. A vertical line is someone stationary, and a person who goes on a journey from A to B has a "world line" which moves through both space and time.

THE RELATIVITY OF SIMULTANEITY

The way someone moves affects the way they perceive the world around them. The key to Einstein's special theory, however, is the knowledge that the speed of light is always the same (pp. 32-33). Any observer, no matter how they move, measures the speed to be c. Einstein explained his argument with trains, moving past a stationary platform.

1 MOTION IS RELATIVE TO THE OBSERVER

Imagine a man is standing in a train that is moving at a steady speed in a straight line. Because motion is relative, he can say the train is at rest, and the platform is moving past him backwards. Inside the carriage, light pulses are sent out in opposite directions from the centre simultaneously. Each beam of light travels at the same speed, c, towards either end of the carriage.

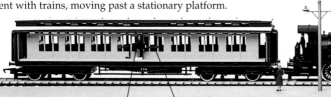

Man sees train as stationary

Light beams emitted in centre of carriage

Platform is moving backwards for man on train

2 MAN SEES SIMULTANEOUS LIGHT BEAMS

The man riding in the carriage will see that each light beam covers the same distance – half the length of the carriage – in the same time. According to the man on the train they reach the two ends simultaneously. Measured from the train, the speed of light is always c relative to the train, whatever speed the train moves at.

Man sees light reach ends of carriage simultaneously

Light beams travel at speed c

Platform has "moved" back in the time taken for light beams to reach the end of carriage

3 WOMAN SEES THINGS DIFFERENTLY

A woman standing on the platform watching the train go by also sees the light pulses travelling in opposite directions at the same speed as each other. Measured from the platform, the speed of light is always c relative to the platform. Because the train is moving, the back end catches up with the pulse going backwards. It meets the end of the carriage before the other pulse catches up with the front.

Woman on platform sees back of train catch up with light beam

Light beams seen by woman do not reach ends of carriage simultaneously

Train has moved forward in time taken by light beam to reach front of carriage

MINKOWSKI'S MATHS

Hermann Minkowski (1864-1909) was one of Einstein's teachers at the Zurich polytechnic, where Einstein took his degree. Minkowski once described student Einstein as a "lazy dog" who "never bothered about mathematics". A few years later, working at the University of Göttingen, Minkowski explained relativity theory more clearly than Einstein himself had done, by introducing the idea of four-dimensional spacetime. This idea of spacetime geometry then became crucial to Einstein's general theory of relativity (pp. 42-45) in 1915.

Cone of light emitted by the Sun in the future

The Earth enters light cone 8 minutes into future

The Sun in the present

Only objects within this cone can send signals to the Sun "now"

Passage of Earth through time

SUNLIGHT'S WORLD LINE

In a Minkowski diagram, light rays are shown as world lines at 45° to the horizontal. Shallower world lines would indicate movement faster than light, which is impossible. The Earth is just over 8 light minutes away from the Sun – it takes light from the Sun 8 minutes to reach the Earth. In this diagram, the Earth moves straight up the page, so that it does not move in space relative to the Sun. The world line of the Sun's light makes a cone. A beam of light from the Sun travels along the edge of the light cone, and only affects the Earth when the Earth has moved 8 and a bit minutes forward in time, to enter the light cone from the beam.

LENGTH IS RELATIVE

Because observers who move relative to one another cannot agree on whether events like the arrival of light pulses at both ends of a carriage occur simultaneously, they cannot agree on other measurements, such as the length of the carriage. This is because they need to use light, which has a fixed finite speed, to make measurements – like measuring the carriage against the platform – that are relative to each other.

1 SYNCHRONIZING CLOCKS

Clocks that are far apart can be synchronized with signals at the speed of light. This is what we do when we set our clocks by the radio time signal, since radio waves travel at the speed of light (p. 33). Imagine there is a row of clocks along the platform, ready to be synchronized by the woman on the platform. At a pre-arranged time she triggers a light beam to set them all at "zero hour". The two clocks that are exactly level with each end of the moving carriage at a certain time are marked. The length of the train is the distance between those two clocks.

Flag marks the end of the carriage

Beam of light sets clocks

All clocks are set at zero hour when they receive the pulse

2 TRAIN GETS SHORTER

The man on the train says that the train is at rest and the platform is moving past. As the platform moves along, it carries some of the clocks towards the synchronizing light beam, and some away, so that the clocks read different times. The markers indicating the two ends of the carriage are raised when the ends are level with the clocks that indicate the "zero hour", but this happens at different actual times for the train. This is why the measured length is less than it ought to be.

Clock shows zero hour at the "wrong" time

Clocks set by light pulse at rear differ

Clocks already zeroed move on

This picture of the train has been enlarged to show the effect more clearly

Moving objects appear to be shrunk in the direction they are moving

Train is shortened, but height and width remain the same

The amount of length contraction is equivalent, in four dimensions, to the amount of time dilation

SHRINKING TRAINS

A train moving close to the speed of light will appear shorter, but just as tall as when it is still. This is not noticeable at low speeds. A car moving at 160 km (100 miles) per hour seems to shrink by 2 trillionths of 1 per cent. In the equations time appears with a minus sign, so when length shrinks, time expands.

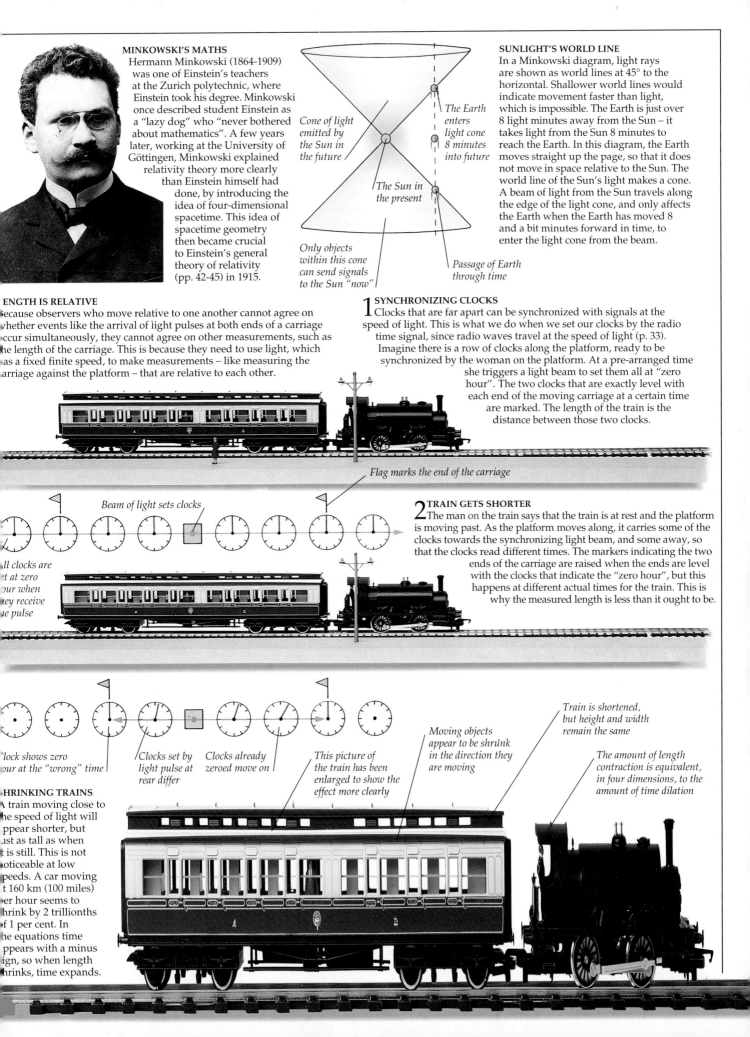

Special theory: proofs and tests

RELATIVELY FAMOUS SCIENTIST

Albert Einstein (1879-1955) is famous for his theories of relativity. He published a paper on his special theory in 1905, and worked on his general theory (pp. 42-45) until 1915. As well as these theories which described how the Universe behaves, he explained why the sky is blue, and helped to create quantum theory (pp. 52-53). His PhD examined the behaviour of particles suspended in a fluid. This study can be applied to droplets of water suspended in clouds, or any other suspension.

THE SPECIAL THEORY OF RELATIVITY makes such strange predictions that many people cannot believe it is really true. They prefer to believe the Newtonian "common sense" view of the world based on everyday experience. They think that Einstein's idea is "just a theory", in the everyday sense that it is a crazy idea that might not be true. In science, the word "theory" is only used for an idea that has been tested and that has passed all its tests. The special theory of relativity has been tested many times. The results of those tests are always that Einstein was right and common sense is wrong. The bits of the special theory that run against common sense only show up at speeds that are a large fraction of the speed of light. Because we never travel at those speeds, those effects are never seen. If they were, then they would be common sense. At slow speeds, the special theory agrees with everyday experience. As Hermann Minkowski once said, "these views of space and time" have "sprung from the soil of experimental physics, and therein lies their strength … henceforth space by itself, and time by itself, are doomed to fade away into mere shadows".

MOVING CLOCKS RUN SLOWLY

One strange prediction of the special theory is that time measured by a moving clock will run more slowly than that of a clock that is stationary. This can be demonstrated by looking at a special kind of clock using light beams, carried on a train that travels at nearly the speed of light. This rule applies to all moving clocks.

1 MAN OBSERVES AT SPEED
Experiments show that the speed of light is the same for everybody, so light makes the ideal clock. Imagine a light beam emitted from a device on the floor of a railway carriage. The light bounces off a mirror in the roof, and back down to a detector further along the carriage. The time it takes for light to travel up and down, at speed c (pp. 32-33), represents one "tick" of the light clock. The man on the train sees the light travel a short distance within the carriage, at speed c.

2 WOMAN SEES A DIFFERENCE
If the train speeds past a woman on the platform, she sees the train move forward while the light beam is bouncing up and down. The distance travelled by the light beam is longer, still at speed c. According to the woman it takes longer for one tick of the light clock on the train.

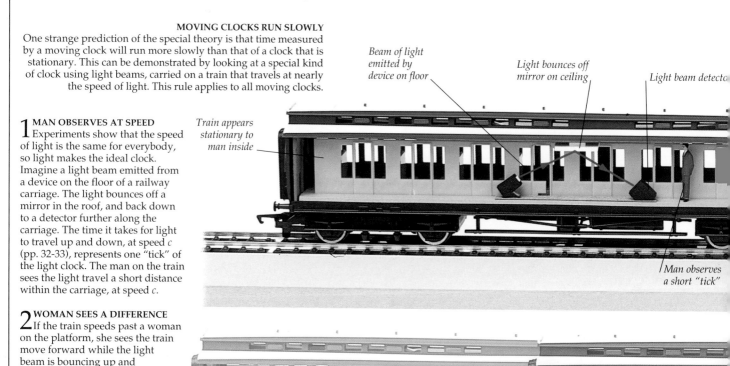

Beam of light emitted by device on floor

Light bounces off mirror on ceiling

Light beam detector

Train appears stationary to man inside

Man observes a short "tick"

Light emitted by device on floor, in train's starting position

Woman sees a long "tick"

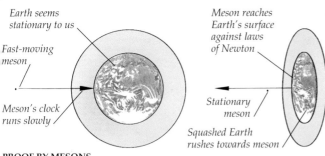

Earth seems
stationary to us

Fast-moving
meson

Meson's clock
runs slowly

Meson reaches
Earth's surface
against laws
of Newton

Stationary
meson

Squashed Earth
rushes towards meson

FLYING CLOCKS

One atomic clock
(p. 56) has been flown round
the world, and the time on it
compared with an identical atomic
clock left on the ground. Concorde, faster
than the speed of sound, can only travel
at about 0.0002 per cent of the speed of light.
Although the time dilation effect is measured in billionths of a
second at these speeds, the atomic clocks are sensitive enough to
show the slowing down of time predicted by Einstein's theory.

PROOF BY MESONS

Particles called mesons are created by the impact of cosmic rays on top of
the atmosphere. Mesons only "live" for a couple of microseconds, not long
enough to reach the ground. They travel so fast that their "clocks" run slowly.
Each of their microseconds is nine microseconds on Earth, so they have
nine times longer to make the journey. For a meson, they are stationary and
Earth is rushing past. It moves so fast that the atmosphere shrinks to one-
ninth of the thickness we measure, so they only have one-ninth as far to travel.
Thus mesons have been detected on Earth, against common-sense logic!

NUCLEAR POWER PROOF

Nuclear power stations, like this
one, generate heat by turning
mass into energy. They do so
by bombarding certain unstable,
or radioactive, heavy atoms
(p. 58) with small particles like
neutrons, so that they split into
two or more lighter atoms, in
a controlled reaction. This is
called fission. The total mass of
all the pieces that such an atom
breaks into is less than the mass
of the original atom. The extra
mass emerges as energy. This is
predicted by the special theory.

EXPLOSIVE PROOF

The special theory says that
mass and energy are related.
They can be interchanged in the
equation $E = mc^2$. Because the
speed of light, c, is so big, a
small mass, m, can be converted
into a huge amount of energy,
E. In 1939 Einstein sent a letter
to the US president Roosevelt,
suggesting that the relationship
between mass and energy could
be used to generate electricity,
or to make a bomb for use as a
deterrent in the Second World
War. He never thought it would
actually be used. The atom
bomb in this photograph was
tested at Bikini atoll in 1946.

PARTICLES ACCELERATE AS PROOF

Particles like electrons (p. 58) can be made to go
very fast in accelerators like this circular one at
Fermilab, Illinois. They are accelerated using
magnetic fields. The more energy the magnetic
fields give the particles, the faster they go. The
extra speed they pick up is never enough to
make them go faster than light, as nothing can
travel that fast. Each extra bit of energy increases
the electron's speed by a bit less. The extra energy
goes into making the particles heavier. These
effects were predicted by Einstein's theory.

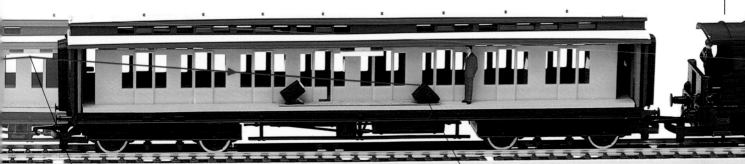

Train has moved forward by the
time the light beam hits the mirror

Train has moved still further by the time
the light reaches the detector on the floor

The philosophy of time

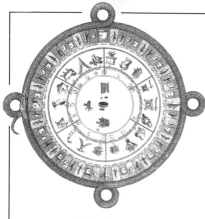

THERE ARE TWO MAIN PHILOSOPHIES of time. One theory suggests that the flow of time from the past to the future is genuine, and only the moment "now" is real at any one time. This seems to mean that we need another "layer" of time in which to measure the "flow" of our time. If our movement through time implies the existence of a "supertime" to measure the rate time passes, there must also be a "super-supertime" to measure supertime, and so on. This idea was proposed by 20th-century British philosopher John William Dunne. The other possibility is that everything that has happened and everything that will happen exists somewhere in four-dimensional spacetime. This means that, in theory, every moment coexists, in a spread-out spacetime reality. All that "moves" is our perception of "now". One version of this has been proposed by the British astronomer Sir Fred Hoyle. The problem is that this seems to leave us with no options for making choices, that is, it means we have no free will.

CYCLIC TIME

The Mayan people in Mexico had a philosophy of time that was based on cycles. They included a 260-day "Sacred Year", a 365-day solar year, and a cycle of 18,980 days, made by 52 solar years, or 73 Sacred Years. Mayans believed history repeated in cycles called "katun", each 7,200 days long, which did not just repeat. Each new katun followed the last cycle that had ended on a day with the same number in the 18,980-day cycle that the new one was beginning. The system is even more complicated than leap years!

A person's early memories are often of favourite toys and games

Not all memories are good; illness may have been a part of someone's life

Most people remember their first car, house, or wage packet

COSMIC SORTING OFFICE

Hoyle suggests that our image of time as an ever-rolling stream is an illusion. Everything that ever was, and ever will be, is "always" laid out in spacetime, just as all the towns in a country are always laid out across the countryside. Imagine all the moments in one person's life arranged in pigeonholes in a cosmic sorting office, and look at the parcel in one pigeonhole at random. Whichever pigeonhole is being examined will seem to be "now", and at that moment the person will have distinct memories of the past, but only vague ideas about the future. Even if pigeonholes are chosen at random, at any moment, or "now", the person remembers the past but knows nothing about the future. We could all live our lives many times over, in all kinds of peculiar jumbled-up orders, but it would still seem that time is flowing from past to future.

The future cannot be predicted, and is represented by shadowy, wrapped packages

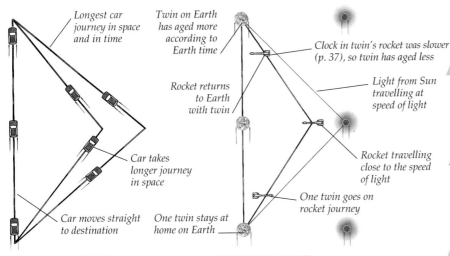

Longest car journey in space and in time

Car takes longer journey in space

Car moves straight to destination

Twin on Earth has aged more according to Earth time

Rocket returns to Earth with twin

One twin stays at home on Earth

Clock in twin's rocket was slower (p. 37), so twin has aged less

Light from Sun travelling at speed of light

Rocket travelling close to the speed of light

One twin goes on rocket journey

SPACE VERSUS SPACETIME

If someone drives a car in a straight line from one place to another, the distance measured on the clock will be less than if the same car travels between the same two places round two sides of a triangle. Things are different in spacetime. When space and time are combined in the special theory, the time component of the equations has a minus sign.

THE TWIN PARADOX

In this Minkowski diagram (p. 34), one of a pair of twins stays at home on Earth, with a world line straight up the page. The other twin takes a trip into space and back on a rocket, making two sides of a triangle in spacetime. The time measured on the Earth clock is more than the time measured on the clock that travels into space and back.

SLICES OF TIME

Ripples spreading out from the point where a pebble has been dropped in a pond show the passage of time. Each moment in time can be thought of as a slice through spacetime – a snapshot of how things are at any instant. Perhaps all the slices are stacked together "all the time", and our perception skips along from one to the other as time passes. This is like the way a movie is made up of successive still pictures.

The present is like a new package, being unwrapped

The immediate future cannot be predicted, but might be anticipated, like a trip planned abroad

COSMIC POSTMAN

Sir Fred Hoyle (born 1915) is one of the greatest astronomers of his generation. He helped to explain how stars turn hydrogen and helium into all the elements found here on the Earth. He has argued against the Big Bang theory (pp. 60-61), and maintains that life originated in space, and not on Earth.

Beyond common sense

THE "EUCLIDEAN" GEOMETRY (p. 26) traditionally learnt in school, including the facts that the angles of a triangle add up precisely to 180° and parallel lines always stay the same distance apart, only applies to flat surfaces. These rules are used in practical applications, like building houses. They match "common sense". In the 19th century mathematicians realized that different kinds of geometry are needed to describe the equivalents of triangles, parallel lines, and so on, on curved surfaces. The new geometry has an immediate use in describing large-scale features of planet Earth, but apart from that, at first it seemed to be a mathematical curiosity with little practical value. Nowadays, ships and aircraft need to take the Earth's curvature into account when planning their routes to be as short as possible. However, when Albert Einstein (p. 36) realized, early in the 20th century, that space itself is bent by the presence of matter, the mathematical tools he needed were ready and waiting for him. The equations that describe black holes, the fate of the Universe, and the way light is bent by the Sun, are those of 19th-century mathematics.

MAN OF CURVES
Karl Gauss (1777-1855) was one of the first people to understand one form of "non-Euclidean" geometry, when he was still very young. The son of a gardener, he was a mathematical prodigy and was presented at the age of 14 at court to the Duke of Brunswick, who then paid for his education. Gauss made contributions to all areas of mathematics, in particular to number theory, and eventually became a professor at the University of Göttingen.

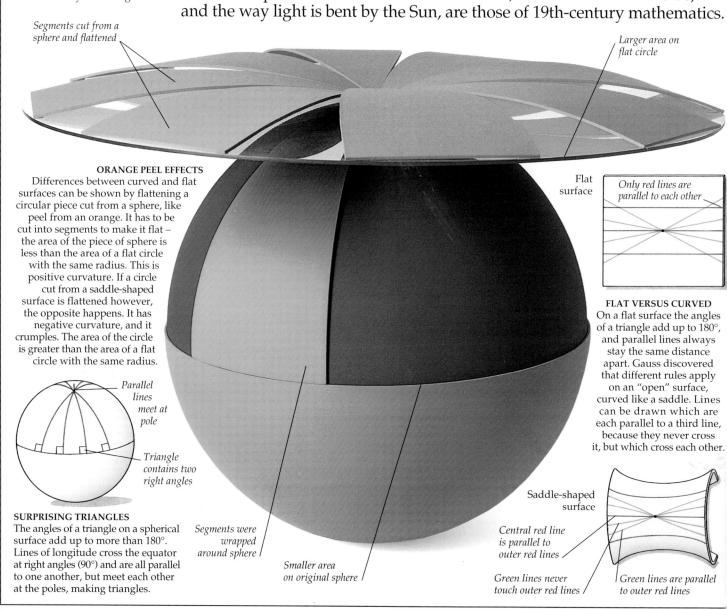

Segments cut from a sphere and flattened

Larger area on flat circle

ORANGE PEEL EFFECTS
Differences between curved and flat surfaces can be shown by flattening a circular piece cut from a sphere, like peel from an orange. It has to be cut into segments to make it flat – the area of the piece of sphere is less than the area of a flat circle with the same radius. This is positive curvature. If a circle cut from a saddle-shaped surface is flattened however, the opposite happens. It has negative curvature, and it crumples. The area of the circle is greater than the area of a flat circle with the same radius.

Flat surface

Only red lines are parallel to each other

FLAT VERSUS CURVED
On a flat surface the angles of a triangle add up to 180°, and parallel lines always stay the same distance apart. Gauss discovered that different rules apply on an "open" surface, curved like a saddle. Lines can be drawn which are each parallel to a third line, because they never cross it, but which cross each other.

Parallel lines meet at pole

Triangle contains two right angles

SURPRISING TRIANGLES
The angles of a triangle on a spherical surface add up to more than 180°. Lines of longitude cross the equator at right angles (90°) and are all parallel to one another, but meet each other at the poles, making triangles.

Segments were wrapped around sphere

Smaller area on original sphere

Saddle-shaped surface

Central red line is parallel to outer red lines

Green lines never touch outer red lines

Green lines are parallel to outer red lines

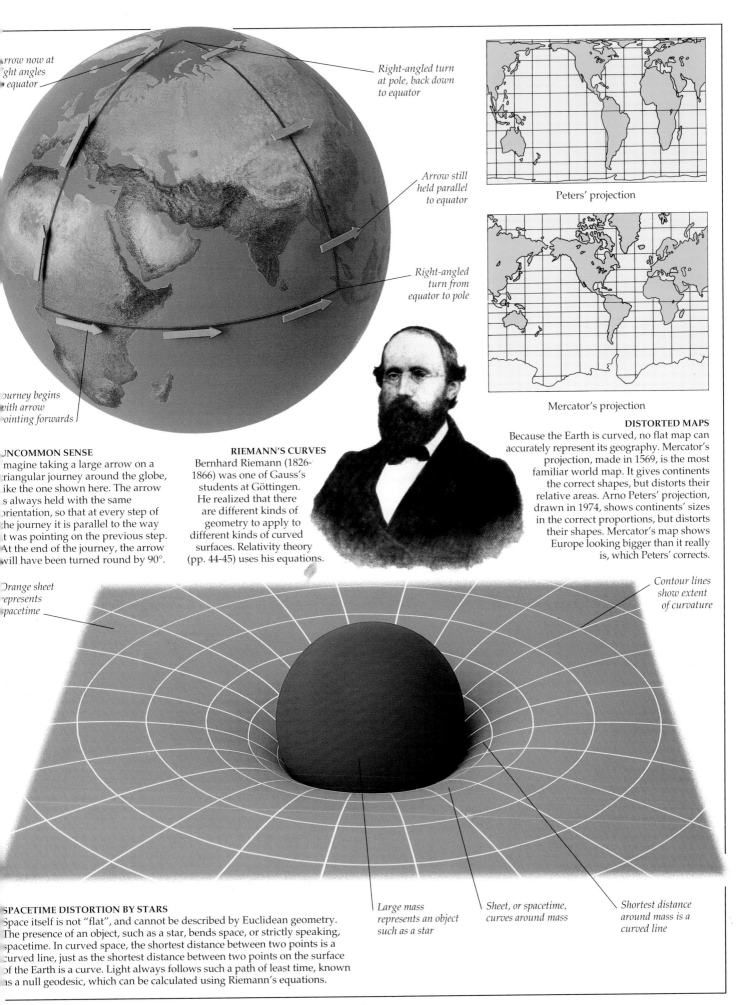

*rrow now at
ight angles
equator*

*Right-angled turn
at pole, back down
to equator*

*Arrow still
held parallel
to equator*

*Right-angled
turn from
equator to pole*

*ourney begins
with arrow
pointing forwards*

Peters' projection

Mercator's projection

UNCOMMON SENSE

magine taking a large arrow on a
riangular journey around the globe,
like the one shown here. The arrow
s always held with the same
orientation, so that at every step of
the journey it is parallel to the way
t was pointing on the previous step.
At the end of the journey, the arrow
will have been turned round by 90°.

RIEMANN'S CURVES

Bernhard Riemann (1826-
1866) was one of Gauss's
students at Göttingen.
He realized that there
are different kinds of
geometry to apply to
different kinds of curved
surfaces. Relativity theory
(pp. 44-45) uses his equations.

DISTORTED MAPS

Because the Earth is curved, no flat map can
accurately represent its geography. Mercator's
projection, made in 1569, is the most
familiar world map. It gives continents
the correct shapes, but distorts their
relative areas. Arno Peters' projection,
drawn in 1974, shows continents' sizes
in the correct proportions, but distorts
their shapes. Mercator's map shows
Europe looking bigger than it really
is, which Peters' corrects.

*Orange sheet
represents
spacetime*

*Contour lines
show extent
of curvature*

SPACETIME DISTORTION BY STARS

Space itself is not "flat", and cannot be described by Euclidean geometry.
The presence of an object, such as a star, bends space, or strictly speaking,
spacetime. In curved space, the shortest distance between two points is a
curved line, just as the shortest distance between two points on the surface
of the Earth is a curve. Light always follows such a path of least time, known
as a null geodesic, which can be calculated using Riemann's equations.

*Large mass
represents an object
such as a star*

*Sheet, or spacetime,
curves around mass*

*Shortest distance
around mass is a
curved line*

Getting a grip on spacetime

ALBERT EINSTEIN (P. 36) EXPLAINED HOW GRAVITY works by imagining that spacetime could be bent by the presence of matter. What we think of as "empty" space is really like an invisible, stretched sheet of rubber. A lump of matter in space – like the Sun – makes a dent in spacetime, like a bowling ball on a trampoline (p. 41). Anything falling past the lump of matter has to follow a curved path, the line of least resistance, called a geodesic. In other words, matter tells spacetime how to bend, and spacetime tells matter how to move. This is Einstein's theory of gravity, the general theory of relativity. Mostly, the way things move is exactly as if there is an inverse square law of gravity. Einstein's general theory of relativity does not overturn Newton's theory (p. 29), but explains why Newton's theory works.

In some special cases Einstein's theory gives slightly different predictions to Newton's theory. In those cases experiments show that Einstein's theory gives the right answers (pp. 44-45). After coming up with his special theory of relativity in 1905 (pp. 34-35), Einstein spent 10 more years, off and on, sorting out the general theory. This was his masterpiece – a complete theory of gravity and, as it turned out, of the Universe.

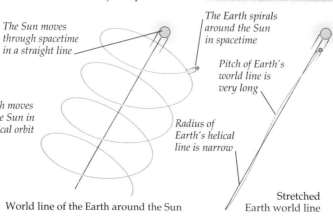

Real position of star

Where we think the star is

FREE FALL
Einstein got the idea for the general theory of relativity when he realized that somebody falling freely does not feel the force of gravity. In orbit around the Earth astronauts are in free fall. These trainee astronauts are in free fall inside an aeroplane, a DC-135, that has a special flying pattern. An apple falling from a tree is in free fall, just as the moon is endlessly falling in its orbit around the Earth (p. 29). "Weightlessness" for an astronaut means feeling that he or she is falling all the time.

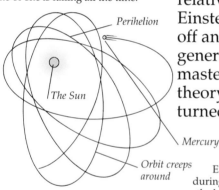

Perihelion

The Sun

Mercury

Orbit creeps around

MERCURY'S WOBBLY ORBIT
The elliptical orbit of Mercury around the Sun acts in a strange way. The whole orbit seems to creep around, or precess. Newton's theory of gravity cannot explain this. Einstein's theory explains it as a dragging effect of the Sun, pulling spacetime near the Sun as it rotates, like a spoon stirring treacle in a tub.

The orbit traces a pattern like a child's drawing of daisy petals. The closest part of the orbit to the Sun is the perihelion, and the effect is called the "precession of the perihelion of Mercury".

BENDING STARLIGHT
Einstein's theory of gravity was tested during an eclipse of the Sun in 1919. With the bright Sun hidden behind the Moon, astronomers could photograph stars in the daytime and measure their positions. Spacetime near the Sun is bent by the Sun's mass, and this model shows the starlight following a bent trajectory, so that the star appears to be slightly shifted. The measured shift was exactly what Einstein's theory had predicted.

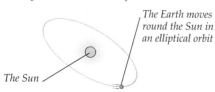

The Sun moves through spacetime in a straight line

The Earth moves round the Sun in an elliptical orbit

The Sun

Space diagram of the Earth

The Earth spirals around the Sun in spacetime

Pitch of Earth's world line is very long

Radius of Earth's helical line is narrow

World line of the Earth around the Sun

Stretched Earth world line

THE EARTH'S WORLD LINE
In space, the Earth follows a closed orbit around the Sun. It moves round the Sun, in a smooth elliptical orbit, once a year. In spacetime, as both the Earth and the Sun move forward in the time dimension the orbit is more like a stretched spring. The Sun spins round, but remains, relatively, in the same place in space, so its world line (p. 34) is a straight line. The Earth's world line stretches the ellipse into a helix around the Sun's world line. In fact, this helical spacetime orbit is so stretched that the "pitch", the distance forward in one revolution, is 63,000 times the radius.

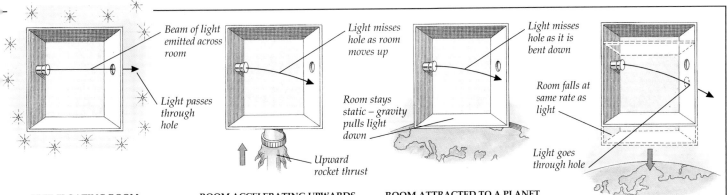

FREE-FLOATING ROOM
Einstein realized that acceleration and gravity are exactly equivalent. In a closed room, floating freely in space too far from any large mass to be affected by gravity, light travels in a straight line across the room, and out through a tiny hole in the wall. If there were astronauts in this room, they would be in free fall.

Beam of light emitted across room

Light passes through hole

ROOM ACCELERATING UPWARDS
If a rocket motor accelerates the room upwards, the imaginary astronauts would fall to the floor. They feel as if they have weight. The room accelerates upwards in the time it takes light to cross it, so the light beam is bent downwards, missing the hole in the wall, as it arrives too low.

Light misses hole as room moves up

Upward rocket thrust

ROOM ATTRACTED TO A PLANET
If the same room is resting on the surface of the Earth, the astronauts would also fall to the floor. Once more, they feel as if they have weight. Gravity pulls the light downwards as it crosses the room, so the light beam misses the hole in the wall, arriving too low. Light is bent downwards by gravity.

Room stays static – gravity pulls light down

ROOM IN GRAVITY FREE FALL
When the same room is falling freely in the grip of the Earth's gravitational field, the astronauts would feel weightless. The light beam is bent by gravity, but just the right amount so that it passes through the hole. It seems to travel straight across the room.

Light misses hole as it is bent down

Room falls at same rate as light

Light goes through hole

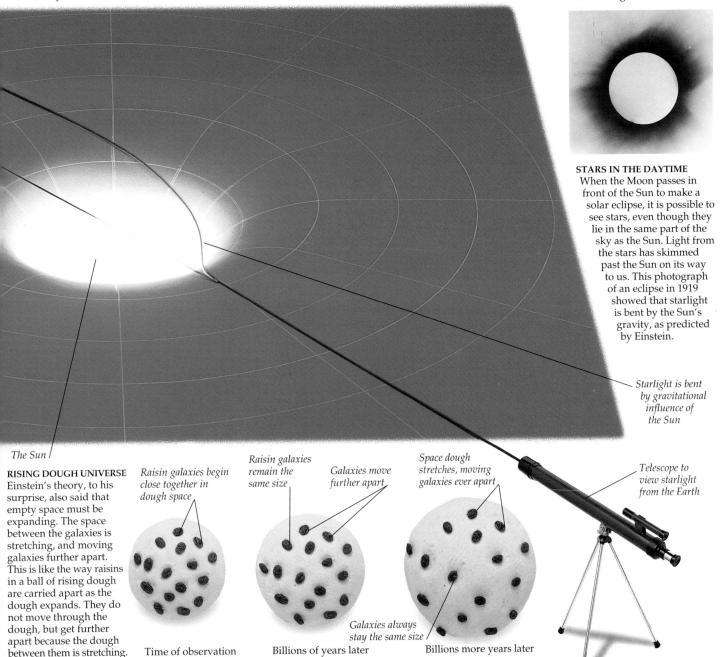

STARS IN THE DAYTIME
When the Moon passes in front of the Sun to make a solar eclipse, it is possible to see stars, even though they lie in the same part of the sky as the Sun. Light from the stars has skimmed past the Sun on its way to us. This photograph of an eclipse in 1919 showed that starlight is bent by the Sun's gravity, as predicted by Einstein.

Starlight is bent by gravitational influence of the Sun

The Sun

RISING DOUGH UNIVERSE
Einstein's theory, to his surprise, also said that empty space must be expanding. The space between the galaxies is stretching, and moving galaxies further apart. This is like the way raisins in a ball of rising dough are carried apart as the dough expands. They do not move through the dough, but get further apart because the dough between them is stretching.

Raisin galaxies begin close together in dough space

Raisin galaxies remain the same size

Galaxies move further apart

Space dough stretches, moving galaxies ever apart

Galaxies always stay the same size

Telescope to view starlight from the Earth

Time of observation

Billions of years later

Billions more years later

General theory: proofs and tests

LIKE THE SPECIAL THEORY OF RELATIVITY (pp. 36-37) the general theory has been tested in many experiments and observations. It has passed every test with flying colours. There is no doubt that Einstein's general theory is the best description we have of the way the Universe at large works. Because the general theory of relativity is a theory of gravity, effects which can only be explained by the general theory show up most clearly where there are strong gravitational fields at work, where there are large masses distorting spacetime. Where gravity is weak, there are only tiny differences between the predictions of Einstein's theory and those of Newton's theory of gravity. Newton's theory is good enough in everyday life. But some experiments are even accurate enough to measure the distortions in spacetime, predicted by the general theory, produced by the presence of the Earth's mass.

SPEEDING CLOCKS
Einstein's theory says that clocks run more slowly in a stronger gravitational field. Gravity gets weaker further from the centre of the Earth. In 1976 an atomic clock was launched on this Scout rocket to an altitude of 10,000 km (6,200 miles), to test it in a weaker gravitational field.

EARTH-BOUND ATOMIC CLOCK
The signals from the clock launched into space in 1976 were compared with the time measured by an identical atomic clock on the ground, at NASA's Wallops Island Flight Center, USA. Here, the gravity space probe is being checked prior to launching the atomic clock. The flight lasted 1 hour, 50 minutes, and the changing speed of the clock on the rocket agreed with the predictions of Einstein's theory to an accuracy of 70 parts per million, or 7 thousandths of 1 per cent.

RIPPLES IN SPACETIME
Two stars in a binary pulsar system (see opposite) make a double "dent" in spacetime. This is similar to the way in which a weightlifter's barbell would distort the surface of a trampoline. If the barbell stayed still, the distorted shape in the trampoline surface would also stay still. If the barbell jiggled about, it would make ripples in the surface of the trampoline. Einstein's theory predicted that objects jiggling about in spacetime would make ripples in spacetime – as gravitational waves.

This light represents a pulsar

Stars make a double "dent" in the spacetime "trampoline"

System is seen by an observer who stays in the same position

MAKING WAVES
The stars in the binary pulsar rotate round one another and make waves in spacetime. Seen from a stationary position, as the pulsar system rotates, fluctuations like waves in the gravitational field are recorded. These gravitational waves carry energy away from the stars. As they lose energy, they slow down. The rate at which the pulsar is slowing in its orbit exactly matches Einstein's calculations of the energy carried away by gravitational radiation.

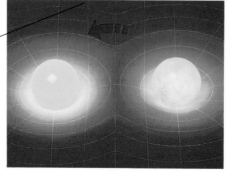

Stars rotate anti-clockwise

Neutron star moves around pulsar

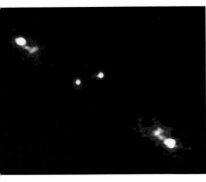

GRAVITATIONAL LENSING EFFECT
When enough mass is gathered in one place it can bend spacetime so much that it acts like a lens bending light. The two images at top left and bottom right in this photograph taken by the Hubble Space Telescope (p. 49) are one and the same very distant galaxy, 10 billion light years away. This galaxy lies behind a concentration of matter that bends the light round two different paths to us. The effect is one predicted by Einstein himself.

Lines of strong gravitational force around star

Emission of waves, eg. radio waves, from the "axes" of star

Star spins round at high speed

WHAT IS A PULSAR?
The most dense concentrations of matter are neutron stars. A neutron star is little bigger than a mountain, but contains as much mass as the Sun. The star has the density of an atomic nucleus (p. 58). Some neutron stars have strong magnetic fields, and beam out radio waves like a lighthouse. They are called pulsars. Changes in the pulsar spin show up as changes in the lighthouse effect.

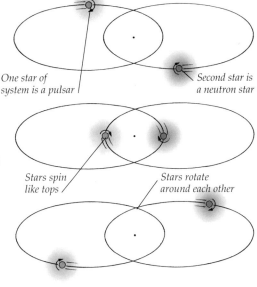

One star of system is a pulsar

Second star is a neutron star

Stars spin like tops

Stars rotate around each other

BINARY SYSTEM OF STARS
In 1974 astronomers discovered a pulsar orbiting around another neutron star. The second star is not a pulsar, but is also made of matter with a density of 500 million tonnes per cubic centimetre (7,400 million tons per cubic inch). The system is called the binary pulsar. The stars orbit each other every 7.75 hours, at a speed of 200 km (125 miles) per second. The distance between them is the same as the radius of our Sun, about 700,000 km (435,000 miles). The period of the orbit increases by 75 millionths of a second each year, in line with the general theory predictions.

This light can represent a neutron star or a black hole (p. 46)

Contour lines are peaks and troughs in gravitational waves

STAR SCIENTIST
Subrahmanyan Chandrasekhar (born 1910) realized in the 1930s, when still a student, that a star which ends its life with a little more mass than our Sun cannot hold itself up against the pull of its own gravity. This was a key step towards the discovery of black holes (p. 46), predicted by the general theory (pp. 42-43). Chandrasekhar also helped to explain how stars work, and came back to black hole studies during the 1970s and 1980s.

Alternating peaks and troughs of gravitational waves appear in any fixed position

INVISIBLE PROOF
Sometimes an ordinary star is orbited by a companion that can only be detected because it emits X-rays. From the way the bright, visible star moves, astronomers can calculate the mass of the companion. If it is more than about three times the mass of our Sun, the X-ray star must be a black hole, as in this system HDE 226868.

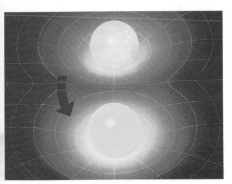

Stars' positions change relative to the observer

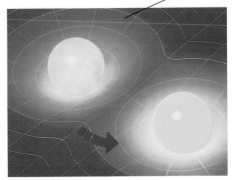

Stars continuously swap places

Black holes and beyond

A BLACK HOLE IS AN OBJECT that has such a strong gravitational pull that nothing, not even light, can escape from it. In the 1780s the British physicist John Michell pointed out that a star 500 times bigger than the Sun but with the same density would trap light in this way. Nobody took the idea seriously until Einstein's general theory of relativity (pp. 42-43) explained gravity in terms of warped spacetime. Einstein's theory says that any amount of matter will wrap spacetime completely around itself, making a black hole, if it is squeezed inside a certain radius. You can make a black hole either by adding more mass to an object and keeping the size the same, or by keeping the mass the same and squeezing it within the critical radius. If the mass is 500 times that of the Sun, it only has to be "squeezed" into a ball about the size of the Solar System. The Sun itself would make a hole in spacetime if it were squeezed down to the size of Mount Everest.

Astronaut

Legs spaghettified on entering black hole

Light moves through spectrum

Ball escapes the Earth's gravitational pull

Ball shot into orbit

Ball with insufficient speed to escape

X MARKS THE SPOT
The core of the nearby spiral galaxy M51 is crossed by a dark "X" of light-absorbing dust, visible in this picture taken from the Hubble Space Telescope (p. 49). The X marks the spot where a black hole, like the ones imagined by John Michell, is at work, eating matter and converting it into energy. It probably contains at least a million times as much mass as our Sun, in a space no bigger than our Solar System. The darker, upright bar of the X may be an edge-on dust ring, 100 light years in diameter.

CRITICAL CALCULATIONS
Just before he died, Karl Schwarzschild (1873-1916) used Einstein's theory of gravity to describe the way spacetime is bent around a lump of matter. He introduced the idea that when a star contracts, there comes a point at which its gravity is so strong that not even light can escape. The critical black hole radius is now known as the Schwarzschild radius. The critical radius is equal to an object's mass multiplied by twice the constant of gravity and divided by the speed of light squared: $2GM/c^2$.

FALLING BACKWARDS
Gravity holds things down on the surface of the Earth and other objects in space. If a ball is thrown straight up into the air, it falls back down. The harder the throw, the higher the ball goes. Similarly, light falls back into a black hole.

GRAVITY AND ORBITS
If a ball could be thrown hard enough, it would go into orbit around the Earth, like Newton's cannonball (p. 28), just as the planets orbit the Sun, and moons orbit planets. An orbiting object is still in the grip of a planet's gravity, although it is in free fall.

ESCAPE VELOCITY
If the ball could be thrown even faster and harder than into orbit, it would escape from the Earth. The escape velocity depends on the strength of gravity of the planet. For a black hole, the escape velocity is greater than the speed of light, but nothing can move that fast.

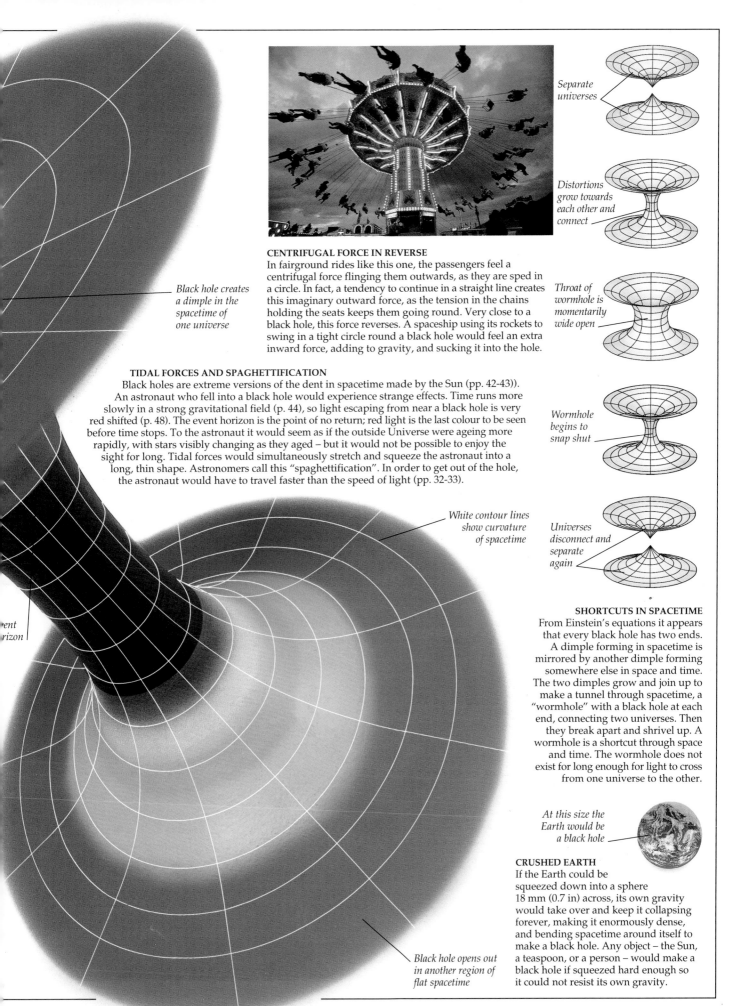

Separate universes

Distortions grow towards each other and connect

Black hole creates a dimple in the spacetime of one universe

Throat of wormhole is momentarily wide open

CENTRIFUGAL FORCE IN REVERSE
In fairground rides like this one, the passengers feel a centrifugal force flinging them outwards, as they are sped in a circle. In fact, a tendency to continue in a straight line creates this imaginary outward force, as the tension in the chains holding the seats keeps them going round. Very close to a black hole, this force reverses. A spaceship using its rockets to swing in a tight circle round a black hole would feel an extra inward force, adding to gravity, and sucking it into the hole.

TIDAL FORCES AND SPAGHETTIFICATION
Black holes are extreme versions of the dent in spacetime made by the Sun (pp. 42-43)). An astronaut who fell into a black hole would experience strange effects. Time runs more slowly in a strong gravitational field (p. 44), so light escaping from near a black hole is very red shifted (p. 48). The event horizon is the point of no return; red light is the last colour to be seen before time stops. To the astronaut it would seem as if the outside Universe were ageing more rapidly, with stars visibly changing as they aged – but it would not be possible to enjoy the sight for long. Tidal forces would simultaneously stretch and squeeze the astronaut into a long, thin shape. Astronomers call this "spaghettification". In order to get out of the hole, the astronaut would have to travel faster than the speed of light (pp. 32-33).

Wormhole begins to snap shut

White contour lines show curvature of spacetime

Universes disconnect and separate again

ent rizon

SHORTCUTS IN SPACETIME
From Einstein's equations it appears that every black hole has two ends. A dimple forming in spacetime is mirrored by another dimple forming somewhere else in space and time. The two dimples grow and join up to make a tunnel through spacetime, a "wormhole" with a black hole at each end, connecting two universes. Then they break apart and shrivel up. A wormhole is a shortcut through space and time. The wormhole does not exist for long enough for light to cross from one universe to the other.

At this size the Earth would be a black hole

CRUSHED EARTH
If the Earth could be squeezed down into a sphere 18 mm (0.7 in) across, its own gravity would take over and keep it collapsing forever, making it enormously dense, and bending spacetime around itself to make a black hole. Any object – the Sun, a teaspoon, or a person – would make a black hole if squeezed hard enough so it could not resist its own gravity.

Black hole opens out in another region of flat spacetime

Across the Universe

THE SUN IS SO FAR AWAY from the Earth that its light takes more than 8 minutes to reach us (p. 35), yet the Sun is very close to us by astronomical standards. The most distant known planet in our Solar System (pp. 14-15), Pluto, is on average 40 times as far from the Sun as we are. Our entire Solar System is just a speck in the cosmos. We are part of the Milky Way galaxy – more than 100 billion stars spread across a disc 100,000 light years across and 1,000 light years thick. One light year is the distance light can travel, at 300,000 km per second (186,000 miles), in a year. Beyond the Milky Way lie many millions of other galaxies. The most distant objects visible in our telescopes are so remote that light from them takes 10 billion years on its journey to the Earth. Astronomers analyse the light from distant galaxies, spreading it out into different wavelengths, or colours, called the spectrum. Dark bands appear in the spectrum, caused by the absorption of some of the light by certain chemicals in the stars' atmospheres. Compared with absorption lines in spectra from the Sun, the bands are shifted towards the longer wavelength end of the spectrum, the red end. The further away the galaxy is, the further the bands move, showing that the galaxies are rushing away from the Earth. This is known as "red shift".

THE WORLD'S LARGEST
Astronomers build their telescopes on the tops of tall mountains, above as much as possible of the Earth's atmosphere, to make the view across the skies much clearer. This is the dome of the Keck telescope, 4,205 m (13,800 ft) above sea level, near the summit of Mauna Kea, in Hawaii. The telescope has a main reflecting mirror 10 m (400 in) across, making it the largest in the world. The segments of the mirror are continually steered and adjusted by computer, 100 times per second to compensate for bends in the segments caused by gravity. Even with a telescope this big, several minutes are needed to obtain the spectrum of a very distant galaxy.

ANDROMEDA RED SHIFT
Because the Universe is expanding (p. 43), light from distant objects is stretched on its journey to us, moving towards the red end of the spectrum and seen as the red shift effect. For the Andromeda galaxy, a near neighbour of the Milky Way, there is no red shift effect.

Absorption line hydrogen H_α

Absorption line hydrogen H_β

Absorption line hydrogen H_γ

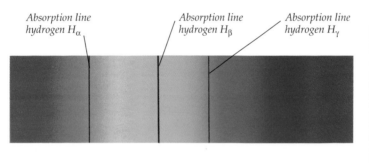

Andromeda galaxy and its spectrum

COMA CLUSTER SHIFT
Galaxies in the Coma cluster, about 350 million light years away, show up as fuzzy blobs in our telescopes. The size of their red shift, 0.022, is proportional to the distance to these galaxies. A galaxy with twice this red shift is twice as far away.

Absorption line hydrogen H_α

Absorption line hydrogen H_β

Absorption line hydrogen H_γ

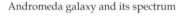

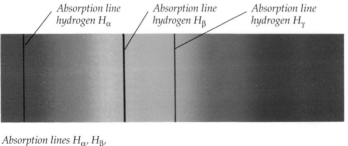

Absorption lines H_α, H_β, and H_γ are shifted beyond the visible spectrum

Coma cluster and its spectrum

4C 41.17 RED SHIFT
Galaxy 4C 41.17 is so distant that it has a red shift of 3.8, so it is about 10 billion light years away. For red shifts bigger than 0.3, the proportional red shift rule no longer works, and astronomers use a more complicated rule based on Einstein's theories.

The light that made this photograph left 4C 41.17 5 billion years before our Solar System was born

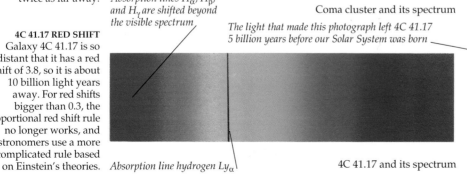

Absorption line hydrogen Ly_α

4C 41.17 and its spectrum

HUBBLE'S UNIVERSE
Edwin Hubble (1889-1953) trained as a lawyer, but soon gave up this career and took up astronomy. He was wounded fighting in France in the First World War. In the 1920s, using the then new 2.5-m (100-in) telescope at Mount Wilson, in California, he showed that many of the fuzzy "nebulae" are other galaxies beyond the Milky Way, and discovered the rule relating red shift to distance, now known as Hubble's Law.

Apparent movement of nearby "test" star

Apparent movement of "test" star across sky

"Test" star slightly further away

Distant stars in background

Nearby "test" star

Angle made by two lines of sight as they cross

Line of sight

Earth moving around Sun

The Sun

The Earth moving around the Sun

DISTANCE BY PARALLAX
Distances to nearby stars are measured (above) by a simple geometrical calculation called "parallax". As the Earth moves around the Sun, the position of a nearby star seems to shift against the background of distant stars. This is like the way a finger held at arm's length seems to shift when you close each of your eyes in turn. The angle the star moves across is its parallax. The bigger the parallax, the closer the star must be. Parallax is measured in parsecs, derived from "parallax second of arc". There are 3,600 seconds in a degree of angle. One parsec is about 3.26 light years.

ORBITING SPACE TELESCOPE
The first large space telescope, named after Edwin Hubble, was launched from the Space Shuttle in 1990. It has a mirror 2.4 m (94 in) across, almost as large as the "100-inch" Hubble used.

"IT'S LIFE, JIM ..."
Light from distant stars and galaxies takes many years to cross space to Earth (p. 24). Similarly, signals, like radio and TV waves travelling at the speed of light (pp. 32-33), diffuse out from the Earth into the cosmos. Any alien civilization that has the technology to receive TV signals, and is 25 light years away from us, will now be receiving broadcasts that were made 25 years ago, like the early episodes of *Star Trek*.

Across the universes

WHEN THE UNIVERSE is faced with "choices" at a basic level, it decides between them at random, in accordance with the laws of probability. This is quantum theory. It also says that the choice is not made until somebody looks to see what is going on. All this is so strange that some physicists have tried to find an alternative explanation, which is just as strange. The only alternative to making quantum choices at random seems to be if all possibilities are "real". This can be demonstrated by an experiment with an electron and two slits, to see which slit the electron goes through (p. 52). Instead of the electron "choosing" which one, the Universe splits in two. In one universe it goes one way, in the other it goes the other way. Both universes are real, but we only experience one of them.

SCHRÖDINGER'S CAT
To show how ridiculous the standard version of quantum theory is, a physicist called Schrödinger dreamed up an experiment involving a cat locked in a box with radioactive material, giving it a 50:50 chance of survival. So what is the state of the cat before the box is opened?

1 DEAD AND ALIVE
Standard quantum theory says that until someone looks in the box, both possibilities of the cat's life remain equally likely, in a "superposition of states". The cat is both dead and alive at the same time. Schrödinger said this was absurd, so the standard theory must be wrong.

Cat dies if poison bottle breaks

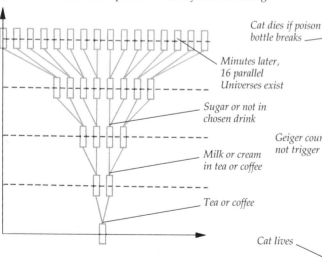

Minutes later, 16 parallel Universes exist

Sugar or not in chosen drink

Milk or cream in tea or coffee

Tea or coffee

Geiger counter does not trigger hammer

Cat lives

2 DEAD OR ALIVE
Standard theory says that as soon as someone opens the box and looks, the superposition of states collapses and one possibility becomes real. We literally do not know what is happening before we look in the box. Most physicists do not worry about what happens when nobody is looking. As long as the equations give the right answers to experiments, they are happy. It still seems strange that things only become real when people look at them.

MANY WORLDS THEORY
The idea that every outcome of every quantum choice really does happen is called the "many worlds" theory. We picture this as a repeated branching from different choices. Suppose someone decides whether to drink tea or coffee. In one universe they choose tea, in the other coffee. More choices follow. Milk or cream in the coffee? Sugar or not in the tea? Again, the universes divide. Every universe is real, but there is no way to communicate between them "next door" in time.

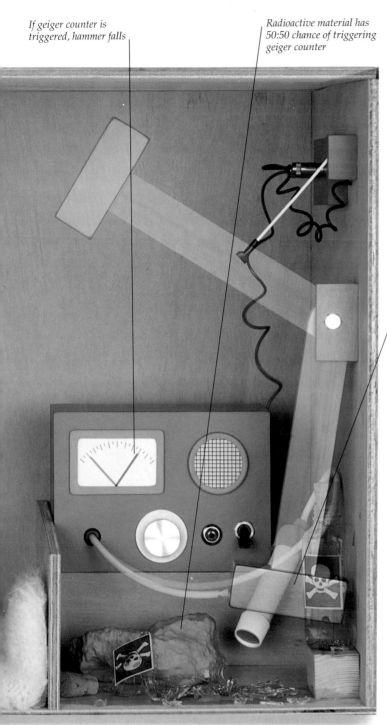

If geiger counter is triggered, hammer falls

Radioactive material has 50:50 chance of triggering geiger counter

Hammer breaks poison bottle

ERWIN SCHRÖDINGER

Erwin Schrödinger (1887-1961) was an Austrian physicist who discovered, in the 1920s, how to describe the quantum world in terms of waves. His theory was known as wave mechanics, and he deduced an equation to describe particle motion. Along with other new theories, it went a long way towards describing previously unexplained phenomena. However, he hated the way his theory became mixed up with ideas of probability and collapsing wave function in standard quantum physics. He said, "I don't like it, and I'm sorry I ever had anything to do with it".

Each path is as valid as the more direct ones

B

A

Direct path between A and B

FEYNMAN'S KNITTING

Newtonian physics (pp. 28-29) says that a particle going from A to B does so along a simple path, or trajectory. Standard quantum physics says it does not follow any path from A to B – it just hops from one to the other. Richard Feynman showed that we can imagine the particle taking every possible path between the two points, all at the same time. Both interpretations give the same answers to quantum calculations. Feynman's "path integrals" explain how an electron passes through two holes at once and makes interference patterns with itself (p. 52).

2 TWIN CATS

If the cat isn't alive on looking in the box, it must be dead. The many worlds theory says that when the radioactive material is given a choice, it takes both possibilities. It triggers the counter, and it also fails to trigger it. The Universe divides into two. In one, the cat is dead, in the other it is alive. In each Universe, someone opens the box to see what has happened. Each person thinks they live in a unique universe, unaware of other worlds.

Bottle of poison is smashed

Geiger counter triggers hammer

Cat dies

ALTERNATIVE REALITY

If the many quantum universes do exist, they are all parallel to each other, and there is no way to get from one to another, except by going backwards in time and then "up" another branch. This idea is common in science fiction. In the *Back to the Future* movies, when Marty goes back in time, he makes a change which alters the future, which was his present. He is really going back down one branch of time and foward up another branch, so that he continues his life in a different quantum universe, never to return to his original world.

God's dice

RICHARD FEYNMAN
Richard Feynman (1918-1988) was one of the greatest theoretical physicists of his generation. He invented a new way to describe particle interactions – Feynman diagrams – and worked out how to combine quantum ideas with electromagnetic theory. He once said that nobody understands quantum theory. Not just a scientist, Feynman played bongo drums professionally, and wrote a best-selling book.

THE IDEA OF A PREDICTABLE "clockwork" Universe (pp. 28-29) was overturned in the 20th century. Physicists discovered that in the case of atoms and subatomic particles (pp. 58-59) the laws of physics actually depend on probability and statistics. If a particle such as an electron can follow more than one path, it is impossible to predict which path it will take. All that quantum theory (pp. 50-51) tells us is the probability that it will go one way or the other. Quantum theory gives us back free will. The reintroduction of uncertainty into the Universe is linked with the discovery that particles behave like waves, and waves behave like particles. In effect, a particle such as an electron is "smeared out" by its waviness. This makes it impossible to say precisely where a particle is, and where it is going – the Universe seems to operate in accordance with the rules of the casino. Einstein hated the idea. He said, "I cannot believe that God plays dice with the Universe", but experiments have proved that Einstein was wrong – the Universe really is ruled by chance. All this uncertainty acting on a small scale, for atoms and smaller particles, means that at least this means our actions are not predetermined in advance.

In-step waves reinforce each other to make "superwaves"

Out of step waves cancel each other out

INTERFERING WITH WATER WAVES
Ripples in water make wave patterns which interfere with one another to make new patterns. If two sets of waves are made, as they spread out and meet each other they interact. Where the peaks of two waves meet, they reinforce each other, and where a peak meets a trough they cancel each other out. The resulting pattern produces lines of "superwaves" and still areas of water. Light waves can be made to interfere with one another by sending them through two small holes. The waves spreading out from each hole interfere and make patterns of light and shade on a screen. This is proof that light is a form of wave. However, other experiments show that light is made up of particles called photons. Light is both wave and particle.

PARTICLES ACTING LIKE WAVES
Feynman's central "mystery" shows up when electrons are made to go through a light interference experiment with two holes. Each electron is fired separately through the slits, and arrives at a TV screen on the other side. The electron makes a single point of light when it hits the TV screen, proving that it behaves like a particle. When a large number of electrons go through the experiment, one after another, they build up a stripey interference pattern of fringes, as if they were waves. Each electron seems to go through both holes at the same time, and undergo interference, before deciding exactly where it belongs in the interference pattern on the TV screen. The electron travels and interferes like a wave, but it arrives as a particle.

Screen with two narrow slits

Electron "waves" split into two

Diffracted electron waves interfere with each other

Waves reinforce each other

Beam of moving electrons

Light fringe on screen made where waves are reinforced

Waves cancel each other out

Dark fringe on screen where waves are cancelled out

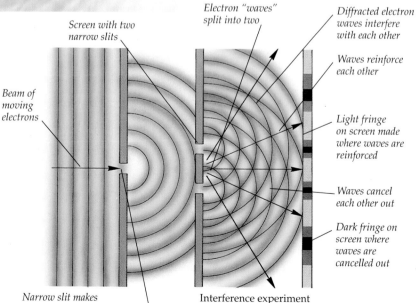

Narrow slit makes electrons spread out, or diffract

Interference experiment with electrons

First few electrons

More electrons

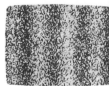

Fringe pattern formed

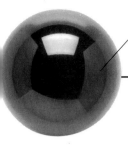

Particles are of opposite charge to cancel each other out

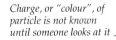
Atom shoots out two particles

Charge, or "colour", of particle is not known until someone looks at it

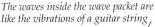

IMAGINARY EXPERIMENTAL BALLS

Suppose an atom shoots out two particles. To balance the equations, they must have "opposite" properties, like charge. The properties can be thought of as one being yellow, and the other red. The rules do not say which particle has which colour in this "thought experiment". There is a 50:50 chance that the one on the left is red, and a 50:50 chance that the one on the right is red.

The waves inside the wave packet are like the vibrations of a guitar string

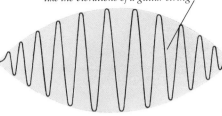

1 BALLS ARE COLOURLESS

Neither particle "decides" which colour it is until it interacts, that is, until someone looks at it. Until the particle interacts, it exists in an uncertain mixture of the two possibilities, a "superposition of states". When someone measures the particle, or looks at it, it "collapses" into the state corresponding to one colour or the other, determining which is which.

QUANTUM WAVE PACKET

Quantum theory tells us, and experiments prove, that subatomic entities do not exist as tiny points of matter. They are spread out, and behave like little wave packets. The wave "collapses" to a point when it interacts with something. When it is on its own, it spreads out more and more. Then, if one bit of the wave interacts, the whole wave collapses down to the point where the interaction happens. Quantum entities are both particle and wave.

Time passes for particle of unknown colour

BELL'S PARTICLE EXPERIMENT

John Bell (1928-1990) was an Irish theorist who thought up the experiment that dashed Einstein's hopes about the laws of probability. He worked out how to test whether particles shot out by atoms really do have properties determined by chance. The experiment was carried out by Alain Aspect's team, in Paris. Aspect's experiment used photons – "particles" of light. The property measured is called polarization. Aspect and Bell proved that when two photons are shot out from an atom together, measuring the polarization of one affects the polarization of the other.

Ball is neither colour until someone looks at one of them

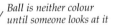
Ball is neither colour until someone looks at it

DOES GOD PLAY DICE?

Quantum rules say what the probability is that an experiment will turn out a certain way. A 90 per cent chance means if the experiment is done many times it will go that way 90 per cent of the time. It is never certain though, and this is like rolling dice. Each number comes up one-sixth of the time if the dice are rolled enough times, but it cannot be predicted which will come up next.

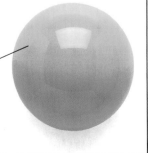

Observer sees that this ball is red

If the other ball is red, this one must be yellow

2 OBSERVER SEES RED

The particle in our "thought experiment" really does not decide what "colour" it is until somebody looks at it. It could equally be red or yellow. As the two particles must have the opposite colours, as soon as anybody looks at one and sees that it is red, the other one must automatically be yellow. This happens instantaneously, without light or any other signal crossing from one particle to the other. Einstein called this "spooky action at a distance". John Bell and Alain Aspect proved that it really does happen.

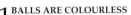

Ultimate length

L IGHT PROVIDES THE ULTIMATE measure of length today. Lengths were originally defined in terms of the human body (pp. 10-11). The Roman mile was made up of 1,000 paces, and the Egyptian cubit was the length of an arm. After the French Revolution, in the 1790s, the French National Assembly established a new system based on the dimensions of the Earth. One metre (1.09 yd) was defined as being one 10-millionth of the distance from the North Pole to the equator, and a "standard metre" rule made of platinum was kept in a vault in Paris, for use as reference. Since 1983, while the basic unit of length is still the metre, it has now been defined as the distance travelled by light in 1/299,792,458 of a second. Several people have suggested redefining the metre so that the speed of light is exactly 300,000,000 metres (186,420 miles) per second (pp. 32-33). Ways of measuring length have progressed from rods or sticks of a standard length to using the speed of light with lasers, infrared beams, and radar, or the speed of sound with sonic measures.

THE ULTIMATE RULER
Light travels in straight lines, so it provides the ultimate ruler in the sense of having a "straight edge". When the two ends of the Channel Tunnel were being bored out from England and France, the digging machines were kept on track with laser beams. Here the red laser beam can be seen making sure that the two ends of the tunnel will eventually meet in the middle.

MAX PLANCK'S QUANTA
Max Planck (1858-1947) discovered in 1900 that energy comes in fundamental units that he called quanta. This was the beginning of the quantum theory of physics. It led to the idea that time and space cannot be subdivided forever.

One millimetre contains 100,000, 000,000,000,000,000,000, 000,000 Planck lengths

PLANCK LENGTH
The fundamental quantum of length is now known as the "Planck length". It is 10^{-35} of a metre, and is the distance light can travel in the "Planck time" (p. 56).

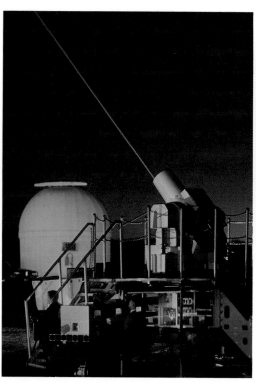

HOW FAR IS IT TO THE MOON?
Apollo astronauts left reflectors on the Moon to bounce back laser beams from Earth. By measuring how long it takes for light to return, astronomers can measure the distance to the Moon to within 30 cm (1 ft). In each month, it varies from 356,500 km (221,500 miles) to 406,800 km (252,800 miles), so the measurement is accurate to one 10-millionth of 1 per cent.

Ship's set path

SONIC TAPE MEASURE

Today, distances can be measured without tape measures or rulers. Because we know the speed of sound, a "sonic" tape measure can be used. This emits a beep of ultrasonic noise, too high for the human ear to hear, which bounces off the wall and back to the instrument. The time taken for the sound to return reveals the distance to the wall.

INFRARED TAPE MEASURE

Infrared beams can measure distances to an accuracy of 2 mm (0.08 in), over a range of 3 km (1.8 miles). This equipment can track moving objects, and also measure the vertical and horizontal angles to features being surveyed. The measurements made are stored in electronic data cards for transfer to a computer to analyse.

Measure is taken to base of instrument

Measure button

Liquid crystal measurement display

Ultrasonic transducer capsules send and receive sonic beams

Electronic "total station"

Prism to reflect infrared beam back

Apparent course of ship, not taking into account factors like tide and wind

Equivalent heading and speed over ground

Factors such as current, wind and tide

"Vector" – projection of next 15 minutes' progress, and "trails" – 15 minutes' history

Closest point of approach to land or vessel, alarm, and time anticipated until then

Latitude and longitude settings locate the ship's whereabouts

RADAR DISPLAY
The ultimate navigation aid today is radar. Radar beams with wavelengths a few centimetres (about 1 in) long are sent out from the radar antenna and bounce off objects, sending back a radio echo that reveals where the objects are. These echoes are converted electronically into pictures on a screen. Sophisticated systems like this ship's navigational radar indicate not only how far away the objects are, but whether they are moving, at what speed, and in which direction.

Collision warning flashes, with audible bleep

Ship on a collision course

Projected path of vessel on collision course

Outline of land mass

This ship appears with its "trail" on its own radar screen

Rings are used to scale distance on ship's course

Touch-sensitive area of screen to change settings

Ultimate time

PHOTO FINISH
Light is the ultimate arbiter in deciding the results of most races, including athletic events, motor racing and horse racing. In close finishes where the naked eye cannot tell, a photo is used to decide which contestant actually finished first. Races are routinely timed to hundredths of a second, and records like running and swimming are given to 0.01 of a second.

ATOMIC TIMERS
Commercially available caesium clocks weigh about 30 kg (66 lb). An individual clock would lose or gain no more than one second every 3 million years. Slightly less accurate, but cheaper, clocks using energy changes in other atoms are widely used by scientists.

LIGHT PROVIDES THE ULTIMATE timekeeper. Time was originally measured by the rotation of the Earth (pp. 16-17) – one rotation is one day of 24 hours, with 60 minutes per hour, and 60 seconds per minute. However, the spin of the Earth is slowing down, so the length of a day is gradually increasing by about 0.0015 of a second each century. Modern clocks can be much more accurate than that. Since 1967, the definition of one second is the time it takes for 9,192,631,770 oscillations of the electromagnetic radiation produced by an energy change taking place in an atom of caesium-133. This radiation is in the same range of wavelength as radio waves in the electromagnetic spectrum (p. 33). The world's time signals are coordinated by the Bureau International de l'Heure, in Paris. It averages time from 80 atomic clocks in 24 countries and sends out signals accurate to one millisecond, defining Coordinated Universal Time, or UTC. Because there is not a whole number of seconds in a year, time signals occasionally include a "leap second" to keep UTC in step with the Earth's rotation.

TIMELY ACCURACY
The ultimate domestic timekeeper is a clock or watch automatically set to UTC every day by a radio signal. The wristwatch on the left picks up radio time signals every 15 minutes using an aerial built in to the strap. A wristwatch so cheap, right, that it may be given away with petrol or some other purchase, will keep time accurately enough for everyday purposes. It is probably as accurate as early chronometers like Harrison's H4 (p. 18), but available at a fraction of their relative cost.

BELL'S FLASHING DISCOVERY
Jocelyn Bell Burnell (born 1943) discovered pulsars (p.45) by accident in 1967. When still a research student with Antony Hewish, she was using a special radio telescope to study distant objects, and noticed a regular flickering signal, keeping perfect time and repeating in just over a second. More than 300 pulsars have since been discovered, some with periods as short as a few milliseconds, others as long as three or four seconds.

One minute contains 600,000,000,000,000, 000,000,000,000,000,000,000,000,000,000 Planck times

PLANCK TIME
Quantum theory says that there is a built-in "graininess" in everything that can be measured, including time. Units cannot be infinitely small, there is a "smallest". It is impossible for there to be an interval of time shorter than 10^{-43} seconds, or a decimal point followed by 42 zeroes and a 1. This fundamental unit is the "Planck time" – the time it would take light to cross the Planck length (p. 54). Both of these units are derived from Planck's constant, a number used in the uncertainty principle (pp. 52-53).

RHYTHMIC FLASHING
A pulsar is a spinning neutron star, with as much matter as our Sun, packed into a ball about 10 km (6 miles) across. It radiates energy in the form of radio waves from a "hot spot" on the surface, like the rotating beam from a lighthouse. The name "pulsar" is a shortened form of "pulsating radio star". It flickers on and off with amazing precision, some hundreds of times per second. One pulsar has a period measured as 0.059029995271 seconds, recorded in September 1974, which changes at a rate of only 0.273 seconds in 1 billion years. Others are the most accurate timekeepers we know – even better than atomic clocks.

Light represents radio hot spot

Star spins round, switching "off"

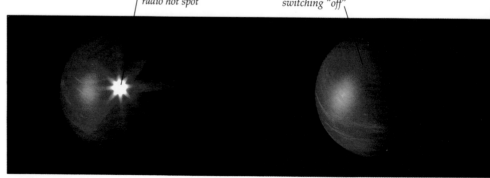

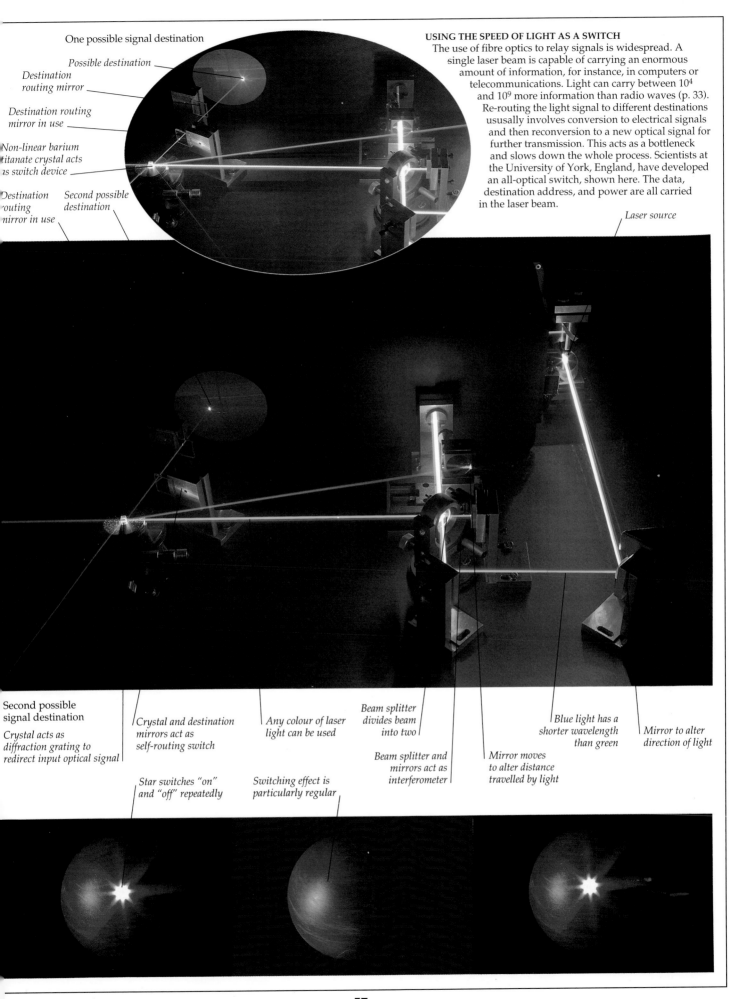

One possible signal destination

Possible destination

Destination
routing mirror

Destination routing
mirror in use

Non-linear barium
titanate crystal acts
as switch device

Destination
routing
mirror in use

Second possible
destination

USING THE SPEED OF LIGHT AS A SWITCH
The use of fibre optics to relay signals is widespread. A
single laser beam is capable of carrying an enormous
amount of information, for instance, in computers or
telecommunications. Light can carry between 10^4
and 10^9 more information than radio waves (p. 33).
Re-routing the light signal to different destinations
ususally involves conversion to electrical signals
and then reconversion to a new optical signal for
further transmission. This acts as a bottleneck
and slows down the whole process. Scientists at
the University of York, England, have developed
an all-optical switch, shown here. The data,
destination address, and power are all carried
in the laser beam.

Laser source

Second possible
signal destination

Crystal acts as
diffraction grating to
redirect input optical signal

Crystal and destination
mirrors act as
self-routing switch

Any colour of laser
light can be used

Beam splitter
divides beam
into two

Beam splitter and
mirrors act as
interferometer

Blue light has a
shorter wavelength
than green

Mirror moves
to alter distance
travelled by light

Mirror to alter
direction of light

Star switches "on"
and "off" repeatedly

Switching effect is
particularly regular

Strings and things

STARTING OUT FROM THE EVERYDAY WORLD, scientists have probed into space in two directions – outwards into the Universe at large, and inwards to the world of the small, atoms and subatomic particles. It is less than 100 years since it was generally accepted that everything is made of atoms – Albert Einstein (p. 36) was one of the scientists who helped to prove this. During the 20th century, scientists have probed within the atom, finding smaller and smaller particles inside. At the very heart of inner space, the smallest entities are thought to be loops and strands of vibrating "string". It would take 100,000,000,000,000,000,000 of these strings, laid end to end, to stretch across the diameter of a proton. All things may be made of strings. To explain the behaviour of larger particles, strings themselves must be made of 10-dimensional pieces of spacetime, rolled up so that only four dimensions are visible. If modern theories are correct, there is nothing "inside" strings. These entities really are the ultimate building blocks of matter, the smallest things in the Universe.

PHILOSOPHER AND POET
The Greek philosopher Empedocles lived from around 490 to 430 BC. As well as being an influential philosopher with ideas about what makes up matter, he was a great poet, and pioneering physician. Later generations thought that he must have been a god in order to achieve so much. However he killed himself by jumping into the volcanic crater of Mount Etna, possibly in an – unsuccessful – attempt to prove that he was a god. All that still remains of his work are 400 lines from one poem, and nearly 100 verses from another poem.

GREEK MATTER
In the 5th century BC, the Greek philosopher Empedocles suggested that all matter is made up of combinations of four elements – earth, fire, air, and water. Earth was said to be dry and cold, fire hot and dry, air hot and wet, and water cold and wet. These four elements were represented by geometrical solids, so that fire is a tetrahedron, earth is a cube, air is an octahedron, and water is an icosahedron. About 400 BC, Democritus suggested matter is made of indivisible pieces he called atoms; but few people took this idea seriously for the next 2,000 years. Physicists, like Isaac Newton (p. 6), accepted the idea of atoms long before chemists were convinced. It was only in the 19th century that the English chemist John Dalton explained that atoms like those of hydrogen and oxygen combine to make molecules, like that of water.

ALL STRUNG OUT
A plucked guitar string (below) can vibrate in several different ways, producing different notes called harmonics. The ends of the string are fixed, but there can also be other places along the string which stay still while the string vibrates. Each stationary point is called a node. The more nodes there are, the higher the note produced. Understanding the nature of the vibrations of string has been useful for theoretical physicists as a basis for research into quarks.

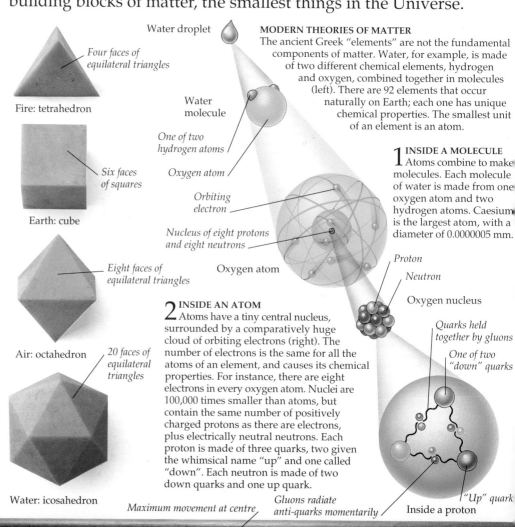

Water droplet

Four faces of equilateral triangles

Fire: tetrahedron

Six faces of squares

Earth: cube

Eight faces of equilateral triangles

Air: octahedron

20 faces of equilateral triangles

Water: icosahedron

Maximum movement at centre

Water molecule

One of two hydrogen atoms

Oxygen atom

Orbiting electron

Nucleus of eight protons and eight neutrons

Oxygen atom

MODERN THEORIES OF MATTER
The ancient Greek "elements" are not the fundamental components of matter. Water, for example, is made of two different chemical elements, hydrogen and oxygen, combined together in molecules (left). There are 92 elements that occur naturally on Earth; each one has unique chemical properties. The smallest unit of an element is an atom.

1 INSIDE A MOLECULE
Atoms combine to make molecules. Each molecule of water is made from one oxygen atom and two hydrogen atoms. Caesium is the largest atom, with a diameter of 0.0000005 mm.

Proton

Neutron

Oxygen nucleus

Quarks held together by gluons

One of two "down" quarks

2 INSIDE AN ATOM
Atoms have a tiny central nucleus, surrounded by a comparatively huge cloud of orbiting electrons (right). The number of electrons is the same for all the atoms of an element, and causes its chemical properties. For instance, there are eight electrons in every oxygen atom. Nuclei are 100,000 times smaller than atoms, but contain the same number of positively charged protons as there are electrons, plus electrically neutral neutrons. Each proton is made of three quarks, two given the whimsical name "up" and one called "down". Each neutron is made of two down quarks and one up quark.

Gluons radiate anti-quarks momentarily

"Up" quark

Inside a proton

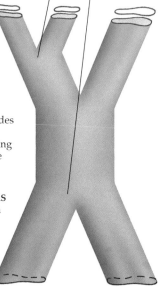

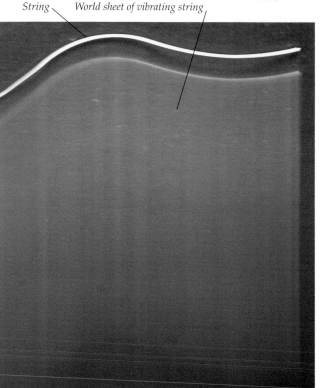

THREE INTO ONE

A hosepipe is made of a sheet of material that has two dimensions, wrapped around in the third dimension. From a distance, it looks like a one-dimensional line. The strings which explain how subatomic particles behave are sheets of 10-dimensional "spacetime", rolled up, or "compactified", so that from a distance only three space, and one time, dimensions are apparent. Strings are so small that even to a quark they "look" four-dimensional.

Sheet of plastic has two dimensions

Sheet is wrapped into a cylinder, to occupy the third space dimension

From a distance, all three dimensions look like one, as a line

Loop of "string"

World sheet of vibrating loop

THEORIST OF EVERYTHING

Oscar Klein (1894-1977) combined gravity, electromagnetism and quantum physics in one theory. It was an early attempt to devise a "theory of everything". Klein used his contemporary Theodor Kaluza's idea that extended Einstein's equations (pp. 42-43) to five dimensions, rather than just four, to explain electromagnetism as well as gravity. However, there are now more forces that need explanation. Klein's theory was the forerunner of string theory, which explains all four forces in terms of 10-dimensional spacetime.

String   *World sheet of vibrating string*

VIBRATING QUARKS

The behaviour of quarks and other particles can be explained by the activities of tiny pieces of vibrating string. Properties such as charge are "tied" to the end of the string. A string moves (up the page) through spacetime, creating a "world sheet" instead of a world line (p. 34). Strings can collide and join up, and then separate. Equations describing this behaviour match the properties of particles measured in accelerators.

A THEORY OF EVERYTHING

Gravity is one of four forces of nature. The others are electromagnetism (p. 33), and two forces, called "strong" and "weak", that operate inside atoms. Scientists trying to understand the other three forces were surprised to find that a string theory which explains them using open strings (left) automatically includes closed loops (above) which explain gravity. Details will take decades to work out, but the theory might eventually explain everything, including the structure of spacetime. It would be the Theory of Everything, or TOE.

Two single loops separate   *Loops combine to make "trousers"*

SPACETIME TROUSERS

When two strings collide, they may join together at their ends to make a third kind of string, which splits apart to make two new strings. If two loops of string combine to form a single loop, they sweep out the world line of a pair of spacetime "trousers".

The birth of space and time

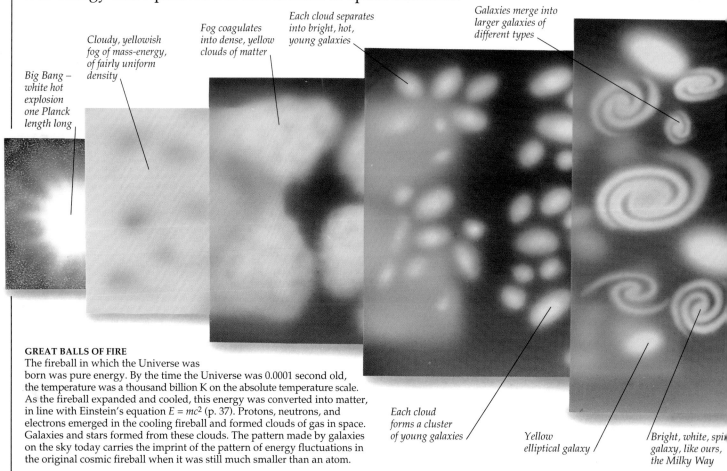

JUST LIKE HOME
Galaxy M51 is like our Milky Way, and this artwork shows how it would look from outside. It is made of hundreds of billions of stars like our Sun. In the Universe, a single galaxy is no more important than a single grain of sand on the beach. Galaxies spread in great chains and filaments across the Universe, like wisps of smoke. These echo the masses of primeval material from which the galaxy chains formed.

THE UNIVERSE WAS BORN IN A BIG BANG. This was not like a big explosion in the middle of empty space. Both space and time actually began with the Big Bang. All the mass and energy filling the Universe today always did fill the Universe, but in the beginning there was very little space to fill. It was a tiny, hot, dense fireball that contained all the energy of the present day Universe. The seed from which the Universe grew was a bubble of spacetime one Planck length (p. 54) across, born with an "age" of one Planck time (p. 56). In the first split second, it expanded rapidly up to the size of a grapefruit, in a process called "inflation". Then, as energy was converted into the kind of matter we know today, mainly hydrogen and helium, the expansion slowed down. The Universe is still expanding, but ever more slowly, 15 billion years later. One theory says that gravity will keep slowing the expanding Universe down, until it eventually stops expanding and starts to collapse back into itself. All the processes of expansion will run backwards, ending in a Big Crunch in which matter is turned back into energy and squeezed out of existence as space contracts.

GAMOW'S HOT AIR
George Gamow (1904-1968) was the first person to work out how hot the Big Bang must have been. Just as air in a bicycle tyre gets hot when it is pumped in, the Universe must have been very hot when it was squeezed into a tiny volume. Gamow said the leftover heat from this fireball should still fill space today, at a temperature of a few K (around -270 °C).

Big Bang – white hot explosion one Planck length long

Cloudy, yellowish fog of mass-energy, of fairly uniform density

Fog coagulates into dense, yellow clouds of matter

Each cloud separates into bright, hot, young galaxies

Galaxies merge into larger galaxies of different types

Each cloud forms a cluster of young galaxies

Yellow elliptical galaxy

Bright, white, spiral galaxy, like ours, the Milky Way

GREAT BALLS OF FIRE
The fireball in which the Universe was born was pure energy. By the time the Universe was 0.0001 second old, the temperature was a thousand billion K on the absolute temperature scale. As the fireball expanded and cooled, this energy was converted into matter, in line with Einstein's equation $E = mc^2$ (p. 37). Protons, neutrons, and electrons emerged in the cooling fireball and formed clouds of gas in space. Galaxies and stars formed from these clouds. The pattern made by galaxies on the sky today carries the imprint of the pattern of energy fluctuations in the original cosmic fireball when it was still much smaller than an atom.

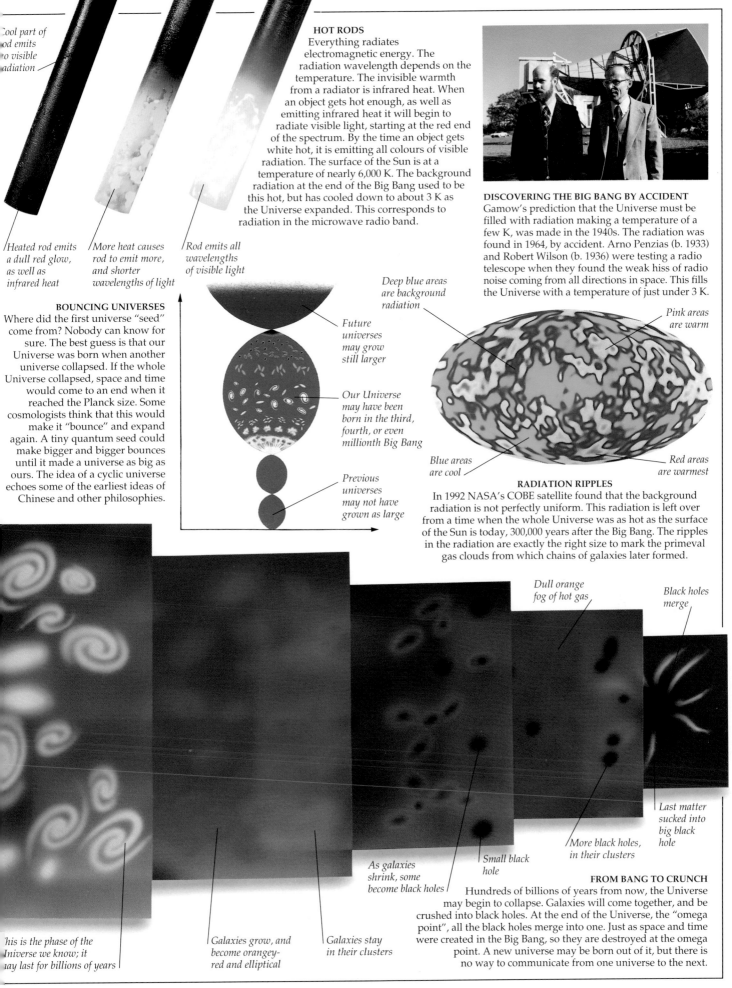

Cool part of rod emits no visible radiation

HOT RODS

Everything radiates electromagnetic energy. The radiation wavelength depends on the temperature. The invisible warmth from a radiator is infrared heat. When an object gets hot enough, as well as emitting infrared heat it will begin to radiate visible light, starting at the red end of the spectrum. By the time an object gets white hot, it is emitting all colours of visible radiation. The surface of the Sun is at a temperature of nearly 6,000 K. The background radiation at the end of the Big Bang used to be this hot, but has cooled down to about 3 K as the Universe expanded. This corresponds to radiation in the microwave radio band.

DISCOVERING THE BIG BANG BY ACCIDENT

Gamow's prediction that the Universe must be filled with radiation making a temperature of a few K, was made in the 1940s. The radiation was found in 1964, by accident. Arno Penzias (b. 1933) and Robert Wilson (b. 1936) were testing a radio telescope when they found the weak hiss of radio noise coming from all directions in space. This fills the Universe with a temperature of just under 3 K.

Heated rod emits a dull red glow, as well as infrared heat

More heat causes rod to emit more, and shorter wavelengths of light

Rod emits all wavelengths of visible light

BOUNCING UNIVERSES

Where did the first universe "seed" come from? Nobody can know for sure. The best guess is that our Universe was born when another universe collapsed. If the whole Universe collapsed, space and time would come to an end when it reached the Planck size. Some cosmologists think that this would make it "bounce" and expand again. A tiny quantum seed could make bigger and bigger bounces until it made a universe as big as ours. The idea of a cyclic universe echoes some of the earliest ideas of Chinese and other philosophies.

Deep blue areas are background radiation

Future universes may grow still larger

Our Universe may have been born in the third, fourth, or even millionth Big Bang

Previous universes may not have grown as large

Pink areas are warm

Blue areas are cool

Red areas are warmest

RADIATION RIPPLES

In 1992 NASA's COBE satellite found that the background radiation is not perfectly uniform. This radiation is left over from a time when the whole Universe was as hot as the surface of the Sun is today, 300,000 years after the Big Bang. The ripples in the radiation are exactly the right size to mark the primeval gas clouds from which chains of galaxies later formed.

Dull orange fog of hot gas

Black holes merge

Last matter sucked into big black hole

As galaxies shrink, some become black holes

Small black hole

More black holes, in their clusters

FROM BANG TO CRUNCH

Hundreds of billions of years from now, the Universe may begin to collapse. Galaxies will come together, and be crushed into black holes. At the end of the Universe, the "omega point", all the black holes merge into one. Just as space and time were created in the Big Bang, so they are destroyed at the omega point. A new universe may be born out of it, but there is no way to communicate from one universe to the next.

This is the phase of the Universe we know; it may last for billions of years

Galaxies grow, and become orangey-red and elliptical

Galaxies stay in their clusters

How to build a time machine

THE MOST SURPRISING IMPLICATION of Einstein's marriage of space and time, the general theory of relativity (pp. 42-43), is that it permits time travel. Einstein realized that the equations describing a black hole have two "ends", so that a black hole has two openings and is a tunnel through spacetime (pp. 46-47). The openings are in different locations in spacetime. The two locations are possibly at different times, not just different places. At first, scientists thought a better understanding of the general theory would rule this out. The idea of tunnels through time and space was not taken seriously (except in science fiction) until the 1980s. Then, relativists decided to prove once and for all that the science fiction writers were wrong and time travel was impossible. To their surprise, they found the opposite – there really is nothing in the laws of physics to prevent time travel, although it would be very difficult to build a working time machine.

BILL AND TED, BACK IN TIME
In the movie *Bill and Ted's Excellent Adventure*, the heroes, seen here visiting Socrates in ancient Greece, use a time and space machine which travels along tunnels, like tangled spaghetti, through spacetime. The general theory of relativity says that the fabric of spacetime might really be like this.

MAKING A TIME AND SPACE MACHINE
To make a black hole big enough to travel through without being spaghettified (p. 46) a lot of mass would have to be gathered together in one place. Such black hole engineering would involve moving stars and planets around, as described in steps 1 to 5 on these pages. The cosmic technology required is far beyond the capabilities of our present civilization, but scientists are intrigued by the possibility because the fact that tunnels through space and time can exist means that they may have occurred naturally. This could help them to explain where the Universe came from.

Stars collapse into a black hole, pulling more stars in with them

Drive your rocket into the wormhole

1 YOU NEED A BLACK HOLE
First, make your black hole (p. 46). To do this, bring a lot of stars together in one place and squeeze them within the critical Schwarzschild radius (p. 47) for the whole heap of stars. We recommend that you use at least 1,000 stars like the Sun to make a hole big enough to work with. Warning: do not do this too close to your home planet, which might get sucked in too!

This end is here and now

This end could be a month in the past – or future

2 POP THROUGH THE WORMHOLE
Drive your spaceship carefully in to the black hole, through the spacetime tunnel, or "wormhole", and out the other end. You will now be in a different place and a different time, say one month ago. Find a large mass – a big planet will do – which you can dangle in front of the wormhole "mouth".

A planet like Jupiter would be a good size to catch

The time and place where you started out

The planet's gravity acts like a magnet for the wormhole

3 PULL THE WORMHOLE ALONG
Dangle the planet in front of the wormhole mouth, and tow it back to where you started. The wormhole mouth will be tugged along behind by gravity, like a donkey following a carrot. Once both ends of the wormhole are near one another, drive round and round in a circle, towing the planet and one end of the hole along.

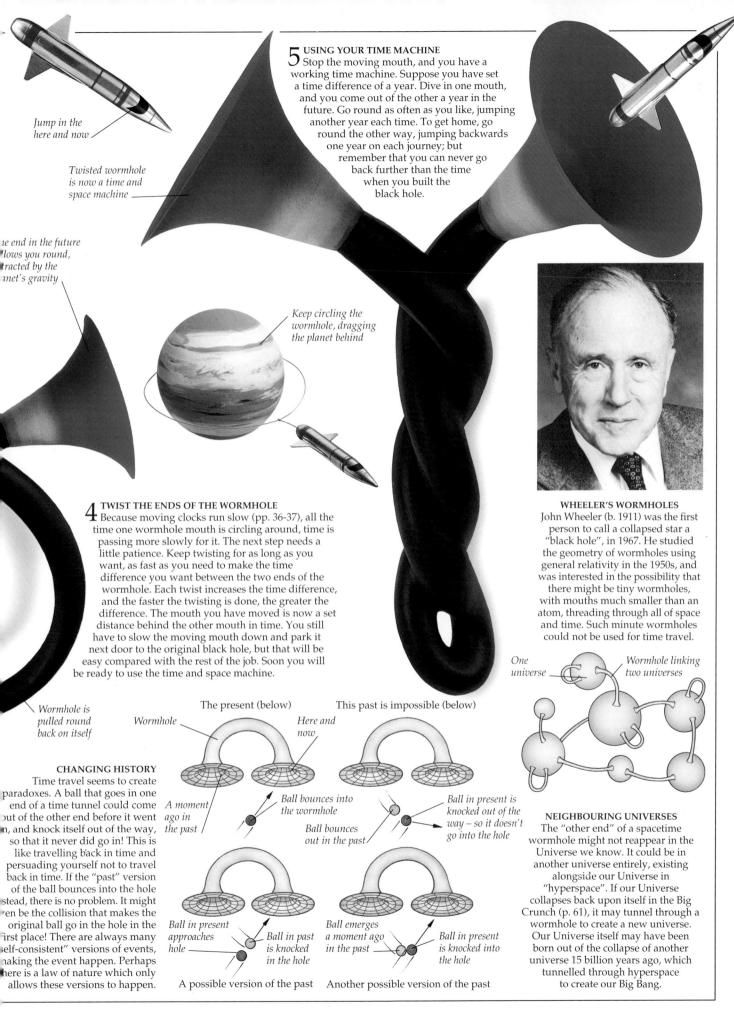

Jump in the here and now

Twisted wormhole is now a time and space machine

...e end in the future ...lows you round, ...tracted by the ...net's gravity

5 USING YOUR TIME MACHINE

Stop the moving mouth, and you have a working time machine. Suppose you have set a time difference of a year. Dive in one mouth, and you come out of the other a year in the future. Go round as often as you like, jumping another year each time. To get home, go round the other way, jumping backwards one year on each journey; but remember that you can never go back further than the time when you built the black hole.

Keep circling the wormhole, dragging the planet behind

4 TWIST THE ENDS OF THE WORMHOLE

Because moving clocks run slow (pp. 36-37), all the time one wormhole mouth is circling around, time is passing more slowly for it. The next step needs a little patience. Keep twisting for as long as you want, as fast as you need to make the time difference you want between the two ends of the wormhole. Each twist increases the time difference, and the faster the twisting is done, the greater the difference. The mouth you have moved is now a set distance behind the other mouth in time. You still have to slow the moving mouth down and park it next door to the original black hole, but that will be easy compared with the rest of the job. Soon you will be ready to use the time and space machine.

Wormhole is pulled round back on itself

CHANGING HISTORY

Time travel seems to create paradoxes. A ball that goes in one end of a time tunnel could come ...ut of the other end before it went ...n, and knock itself out of the way, so that it never did go in! This is like travelling back in time and persuading yourself not to travel back in time. If the "past" version of the ball bounces into the hole ...stead, there is no problem. It might ...en be the collision that makes the ...original ball go in the hole in the ...first place! There are always many ...elf-consistent" versions of events, ...making the event happen. Perhaps ...here is a law of nature which only ...allows these versions to happen.

WHEELER'S WORMHOLES

John Wheeler (b. 1911) was the first person to call a collapsed star a "black hole", in 1967. He studied the geometry of wormholes using general relativity in the 1950s, and was interested in the possibility that there might be tiny wormholes, with mouths much smaller than an atom, threading through all of space and time. Such minute wormholes could not be used for time travel.

One universe

Wormhole linking two universes

NEIGHBOURING UNIVERSES

The "other end" of a spacetime wormhole might not reappear in the Universe we know. It could be in another universe entirely, existing alongside our Universe in "hyperspace". If our Universe collapses back upon itself in the Big Crunch (p. 61), it may tunnel through a wormhole to create a new universe. Our Universe itself may have been born out of the collapse of another universe 15 billion years ago, which tunnelled through hyperspace to create our Big Bang.

The present (below)

This past is impossible (below)

Wormhole

Here and now

A moment ago in the past

Ball bounces into the wormhole

Ball bounces out in the past

Ball in present is knocked out of the way – so it doesn't go into the hole

Ball in present approaches hole

Ball in past is knocked in the hole

Ball emerges a moment ago in the past

Ball in present is knocked into the hole

A possible version of the past

Another possible version of the past

Index

Acknowledgments

Dorling Kindersley would like to thank:
Selfridges Ltd, London, for loan of telescope p. 43br; Lillywhites Ltd, London, for loan of snooker balls p. 53l, 53r; Hall & Watts Optronics Ltd, St Albans, for photograph of electronic land surveying equipment p. 55tr; Prof. J. K. A. Everard of University of York for permission to photograph his laser switch device, which he developed while at King's College, London, p. 57t; Geoffrey Wheeler for loan of surveying equipment p. 6bc, 6br; Capt. Kenneth A. Long FNI, Maria Blyzinsky for consultancy advice; Mark Boon, Richard Clemson, Eric of Sue Boorman Animal Promotions for modelling; Wendy Chandler, Donks Models, Peter Griffiths, Alan Jones for models; Stephen Bull, Helen Diplock, Roy Flooks, Mick Gillah, Janos Marffy, Richard Ward, John Woodcock, Dan Wright for additional illustration; Andy Walker for computer images;

Paul Bricknell, Jane Burton, Andy Crawford, Christi Graham, Steve Gorton, Frank Greenaway, Colin Keates, Dave King, David Murray, Tim Ridley, James Stevenson, Clive Streeter, Harry Taylor for additional photography; Bob Gordon, Johnny Pau for design assistance; Jonny Gribbin for DTP assistance; Sharon Southren for additional picture research; Sharon Jacobs for proofreading.

Illustrations Nick Hall
Photography Tina Chambers
Index Jane Parker

Publisher's note No animal has been injured or in any way harmed during the preparation of this book.

Picture credits

t=top b=bottom c=centre l=left r=right

American Institute of Physics/Emilio Segre Visual Archives 59cl, 60tr, 63cl.

AT & T Archives 61tr.
Bildarchiv Preussischerkulturbesitz 35tl, 51tr.
Trustees of the British Museum 11tl, 11tc, 11cr, 11bl, 17tl.
Bettmann Archive 41c, 45cr, /UPI 49tl.
BIMP 11br.
Bridgeman Art Library 10bl, 24tl.
British Airways/Adrian Meredith 37tr.
Jean Loup Charmet 17br, 20tl.
CERN/Anita Corbin and John O'Grady /The British Council 53cl.
Bruce Coleman 20cl.
English Heritage/Skyscan 26cl.
E. T. Archive 8bl.
Mary Evans Picture Library 14tl, 25tl, 28tl, 38tl, 40tl, 46bl, 58tl.
Werner Foreman Archive 27tl.
Robert Harding Picture Library 9tl, 9cr, 10tl, 26bc, 29cr, 30tl, 37cl.
Michael Holford 11bl, 29br.
Hulton Deutsch 24bc, 31cl, 36tl, 39br.
Hulton/Bettmann 20cr, 30-1, 52tl.
Image Select (Ann Ronan) 6tr, 10cr, 12tl, 16tr, 33bl.
The Independent 17cl.

Kobal Collection 49br, /Amblin Entertainment 51br, 62tl.
Lombard North Central PLC 32c.
Courtesy of the London Transport Museum 7bl.
Mansell Collection 8c, 9bl, 22tl, 26tl, 27l.
NASA 42tl, 44tl, 44cl, 46cl, 46tr, 48cbr, 61cr.
Michigan Technological University/ Dana L. Richter 23bc, 23c.
National Maritime Museum 12c, 13bl, 18bl, 19cl, 19bl.
Open University 56bl.
O.S.F./Paul Franklin 23br.
Q.A. Photos Ltd 54tr.
Racecourse Technical Services 56tl.
Royal Greenwich Observatory 43cr.
Science Museum 19rc.
Science Photo Library 7tl, 7c, 12br, 15br, 20bc, /Pekka Perviainen 27c, /NASA 31tl, /NASA 32tl, 33tl, 37cr, /NASA 45tl, 45br, 48tr, 48cr, 60cr.
Tony Stone Images 13tc.
Ullstein Bilderdeinst 54cl.
Zefa/Bert Leidman 11cl, 29tl, 37cl, 47tl, 54cr.

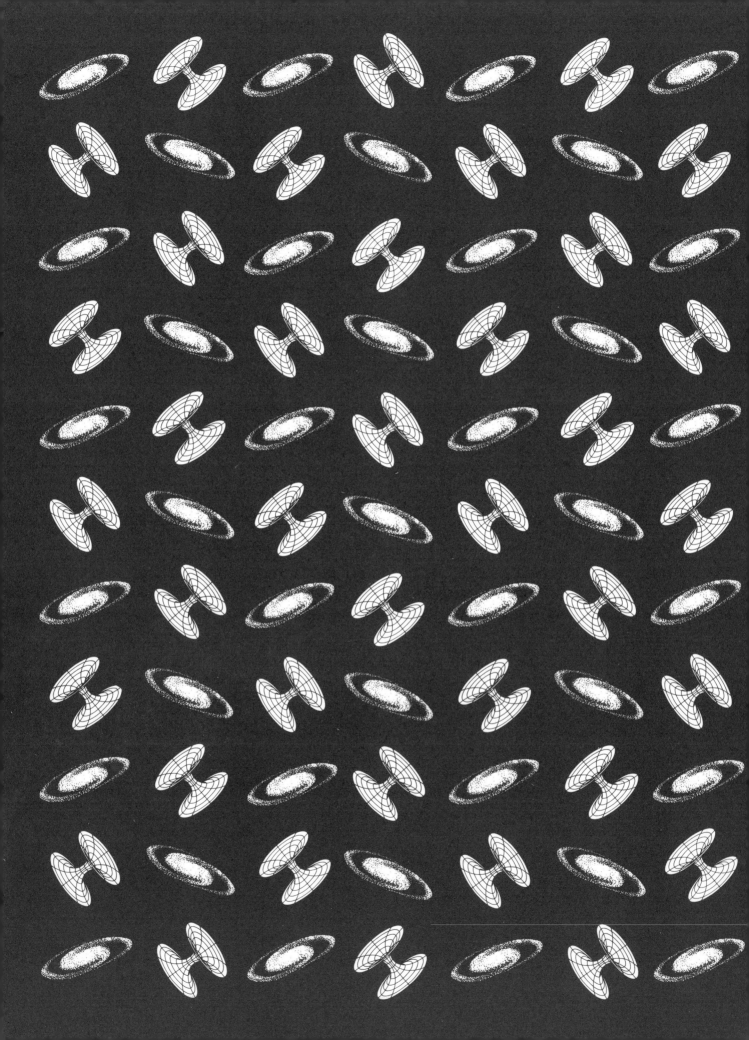